AF531269

DYNAMICS OF RURAL DEVELOPMENT

DYNAMICS OF RURAL DEVELOPMENT

Edited by

Dr. S.N. Tripathy

DISCOVERY PUBLISHING HOUSE

NEW DELHI-110002

First Published-2007
Reprint : 2012
ISBN 978-81-8356-215-7

Published by

DISCOVERY PUBLISHING HOUSE

4831/24, Ansari Road, Prahlad Street,
Darya Ganj, New Delhi-110002 (India)
Phone: 23279245 • Fax: 91-11-23253475
E-mail:dphbooks@rediffmail.com
dphtemp@indiatimes.com

Printed at:

Dynamic Printers

Preface

Rural development and poverty reduction have been important objectives of development policy since the inception of planning in India. Various anti-poverty, employment generation and basic services programmes have been in operation for decades in India. The ongoing reforms attach great importance to removal of poverty, and addressing specifically the wide variations across States and the rural-urban divide. Anti-poverty strategy has three broad components: promotion of economic growth; promotion of human development; and targeted programmes of poverty alleviation to address multi-dimensional nature of poverty. The various programmes targeted at the poor have been streamlined and strengthened in recent years, including through the NREGS.

Incidence of poverty has been estimated by the Planning Commission on the basis of the large scale quinquennial sample surveys on household consumer expenditure conducted by the National Sample Survey Organization (NSSO). The latest such survey relates to the 55th round conducted in 1999-2000, and no official estimates of poverty are available beyond this date. Economic Surveys for 2003-2004 and 2004-2005, on the basis of results of the 55th round, had indicated that there has been an impressive decline in the incidence of poverty in the 1990s. However, the extent of the actual decline in the proportion below the poverty line (BPL) between 1993-99 and 1999-2000 has been a subject of an intense debate by academicians because of the change in methodology for collection of basic data in 1999-2000 and possible non comparability with earlier rounds of the consumer expenditure surveys.

The Food and Agriculture Organization (FAO) advocates per capita per day availability of food-intake has been considered as a robust indicator to measure food insecurity. But in recent years, the anthropometric indicators are increasingly being used to measure the incidence of food-insecurity. Institutions like WHO and UNICEF are largely in favour of anthropometric indicators to measure food insecurity.

Among the poorest segments of the population, increasing their income, their preference towards consuming calorie-intensive foods and the future changes in the relative price of the cereals can only reduce calorie malnutrition. Calorie income elasticity in nearly 0.5 implying that a one per cent change in per capita income increases calorie consumption by half of one per cent. It may take several years for the poorest population experiencing a deficit of calories to reach the required level of calorie consumption. Simultaneously, marginally and non-deficient income groups may also increase their consumption and reach levels that can be substantially higher than requirements. Hence, malnutrition in the poorest groups may co-exist with a calorie surplus. Thus, regular income growth in the poorest population and accelerated food production are unfeasible solutions. The solution may be that reshuffle food availability from the groups with calorie surplus to those in deficit. Malnutrition may be more concentrated in the poorest segment and may be more children specific given the family size and income. Malnutrition is widely prevalent in children due to the fact that calorie-income elasticity for children is lower. Secondly, due to urbanisation women's participation in the labour force has increased resulting in the deterioration of breast-feeding.

Additional food purchased for the infant, out of the additional earnings of the female participating in the labour force, does not compensate for the losses in breast milk consumption; the net effect in deterioration in the infant's nutritional status.

Other factors may also adversely affect the children. Other competing expenditures like transportation, clothing, housing etc. become more important in urban areas, affecting the share of income spent on food. Relative prices of food tend to rise. New commodities become available, such as entertainment and fashion, to be consumed basically by the adults. The result is that the share of income spent on food for children can decline because expenditures other than food become necessary and new commodities, relevant basically to adults, become available.

Calorie malnutrition in young children results from the fact that the family as a whole is malnourished. Secondly, calorie-wise children are discriminated relative to adults. Thirdly, discrimination among children may be manifested according to some characteristics like intelligence, sex, birth order, feeding practices etc. In case, the entire family is malnourished, increase the household overall purchasing power of food-either by increasing their disposable income or lowering the relative price of calorie intensive food. Where there is discrimination among children, preferences of adults in allocation food to particular children can be influenced by emphasising that as an investment good. The rate of return to investment in education is a function of the initial level of nutrition.

India has undoubtedly made considerable progress in the last two decades in promoting child survival and basic education. Along with a supportive policy framework, including the recent legislation to make elementary education a fundamental right, there have been in existence some major initiatives for children. More prominently, these are in the form of centrally sponsored schemes such as the Reproductive Child Health Scheme (RCH) of the Department of Health and Family Welfare. The Integrated Child Development Services (ICDS) of the Department of Women and Child Development and several primary education schemes, including the District Primary Education Programme (DPEP) and the government's flagship programme, the Sarva Shiksha Abhiyan (SSA). As a result, noticeable progress has been made over the last fifty years, as seen in the rapid expansion of the ICDS and primary education delivery network across the country.

Programmes can be effective under the right conditions, and these tend to be related to fundamental causes of malnutrition, such as social exclusion and women's status. Policy decisions and implementation may affect these conditions. In the process of negotiating investment from international financial institutions in nutritional improvement, the basic requirements underpinning programmes may prove feasible issues for modification. At the same time, the simple but crucial decisions to actually commit resources and implement the programmes themselves must be taken. Beyond context, the coexistence of key programmes must be fostered.

Poverty alleviation facilitates nutrition improvements in two ways. First, the programmes themselves provide household resources for improving nutrition. Second, poverty alleviation programme organisation can provide for the structure, outreach, and individual contact that nutrition activities can build on. The poverty alleviation organisations provide the best prospect of extending and supporting local programmes. The direct effects of improvements such as in education, particularly for women, and in water and sanitation, have far-reaching effect on nutrition.

A number of issues that effect all potential programmes remain to be resolved,. Most involve research that should be undertaken at an early stage, and represent sensible investment in terms of programme effectiveness in the future. Some of the salient issues are outlined here.

An issue which has concerned us is defining the contextual factors as well as how these can be changed through policy measures in order to improve programme success. These issues can be researched and guidance provided for programme design. More specifically, knowledge of the relative roles and cost-effectiveness of different community-based and facility-based activities is not yet well laid out. This could be improved from compilation and synthesis of current knowledge. Such research, although of general application, would be most useful in advance of detailed programme planning. The relative effects of using resources for supplementary, feeding, compared with using the same resources for behavioural change, could also be worked out as a generic issue. Population growth in India is in the cities and towns.

As a result, the malnourished are increasingly in urban areas. Yet most planning for improving nutrition refers to villages and rural communities. Research is needed on the most effective means of improving nutrition in urban areas, and where these have common features and problems, this research can be generic. We need to focus on nutrition-relevant programmes in cities and their evaluations, as well as understand better the parallels with rural programmes, especially regarding the appropriate forms of local organisation. This review leads to extensive proposals for finance, and even more for recruiting and training millions of people. Better nutrition and development for children is a powerful outcome objective, with implications for society and development overall. In particular, problems of social exclusion, and discrimination by class and gender, would have some counteracting forces.

For individuals, specific nutritional problems would be well on the way to solution. Micronutrient deficiencies would be becoming a thing of the past, with goiter and iodine deficiency vanishing, and Vitamin A deficiency no longer visible and sub-clinically much reduced. Anaemia would likely be in retreat, with successful fortification. Pregnant women in particular, others too, would be getting regular multiple micronutrient supplements. Low birth weight and early childhood growth failure, and associated sickness and death, would be much reduced—the next generation would be larger, healthier, benefit more from education, and be more productive.

In the light of the above analysis the present work "Dynamics of Rural development" is an attempt to through light on various dimensions in the sphere of economic development through the writings of 15 papers by eminent scholars in the field. The editor expresses his thanks to all the paper contributors, whose papers have enriched the work. Certainly, the present work will be of immense to the policy makers, planners, researchers and general readers.

Dr. S.N. Tripathy

Contents

1

Dynamics of Tribal Development Through Lanjia Saora Development Agency (LSDA)

*Dr. S.N. Tripathy**

Orissa has the distinction of having a high concentration of scheduled tribe population, which according to 1991 Census was 70.32 lakhs constituting 22.2 per cent of the total population of the state. Next only to Chhatisgarh, the state has the largest tribal population in the country accommodating as many as 62 tribal groups.

The Saoras or Savaras or Lanjia Saoras are one of the oldest known tribes of India inhabiting in the hills of Orissa, Bihar, Madhya Pradesh and Maharashtra.

Saoras/Lanjia Saoras have been an extremely backward ethnic group dwelling in specific pockets of Koraput and Gajapati district of Orissa. They have a distinct socio-cultural trait that has been retained due to their inaccessible abodes situated in extremely difficult terrain. Lack of education, skill unapproachable hostile terrains, socio-economic backwardness, poverty and non-responsive to various development programmes pose serious problems in inducing rapid socio-economic development among them. Before focusing light on Lanjia Saora tribe of Orissa, it is imperative to discuss briefly the habitat of tribal areas and their impediments.

* **Principal Investigator, the Major Research Project and Reader in Economics, Aska Science College, Aska (Orissa) 761 111.**

Features of Tribal Areas

The tribal areas have some peculiar characteristics, located in unapproachable remote tracts quite often isolated from the rest of the world. The main features are as follows:

(i) The areas are remotely situated without having the essential infrastructural facilities;

(ii) Those have good natural and physical resources, which have remained unexploited due to non-availability of skill and entrepreneurship among the tribals;

(iii) The backwardness of the areas has a contrasting feature of poverty amongst plenty due to low level of education and techno-economic profile of the tribals;

(iv) Due to ignorance and innocence of tribals they are prone to severe socio-economic of tribals they are prone to severe socio-economic exploitation by non-tribals in several ways;

(v) Forests and tribals go together and quite often, excess dependence of tribals on the local forests creates problems of ecological and environmental degradation.

Development of Tribals and Tribal Areas

Due to extreme socio-economic backwardness, isolated living, because of shyness, ignorance and innocence, inapproachability of the habitats and remoteness of the areas, the development of tribals has been a challenging task.

Article 46 of the Constitution of India enjoins upon the state to promote the social and educational interest of weaker sections particularly of scheduled castes and scheduled tribes to give them adequate protection against social injustice and exploitation.

The strategy for Tribal Development was modified during Fifth Plan when Tribal sub plan approach was introduced, as the more Area Development Approach without specific allocation of funds did not achieve the desired result. A more comprehensive plan for economic and human resource development in tribal areas was necessary for which allocation of funds for tribal development was considered as the imperative need.

Problems of Tribal Development

Development of these primitive tribes in particular has posed serious problems because of several factors:

(i) Their location in unapproachable areas is the main reason for which the benefits of the normal development programme of the tribal development programme do not reach them at all;

(ii) Due to lack of roads and communication facilities the areas are almost shielded and it has been very difficult to carry the benefits of any programme or even assistance of any type to these people;

(iii) Some of these groups repel outsides and become violent at the sight of strangers, which is a discouraging factor in introducing appropriate development programmes in the area;

(iv) These primitive tribes are, by and large, non-cooperative and non-responsive to various social, economic or welfare programmes because of ignorance, simplicity and absence of education or skill among them.

Under these constraints, it was not possible a faster growth either socially or economically among these primitive tribes even under the Tribal sub-plan programmes through the effort of I.T.D.A.s. Besides, some of these primitive tribes are also residing in inaccessible areas even outside the I.T.D.A.s where the impact of normal development programmes could never be felt.

Realizing the shortcomings of the normal programmes and their failure in creating impact on these primitive tribes, the development strategy for primitive tribes was changed and special efforts were initiated by the end of the Fifth Plan by bringing such areas inhabited by the primitive tribes under Micro project activities. The primitive tribes were identified and the areas of their concentration were carved out to be put under a Micro project not only for development of these people through special efforts but also for all sided development of the areas. In course of time, more areas were identified and more micro-projects were set up. The end of 1990-91 has set up 15 such micro-projects up in the Orissa covering 12 primitive tribes. Some of these Micro projects have also been located outside Tribal sub-plan areas.

Lanjia Saora Development Agency (LSDA)

Lanjia Saora Development Agency was set up at Serang in March 1979 for putting special efforts in all sided development of Lanjia Saoras who reside in a compact area near Serang. The Micro Project is functioning since then continuously making a judicious mix of sectoral programmes, infrastructural programmes and beneficiary oriented programmes not only to benefit the Lanjia Saoras in particular but also to develop the agency area in general.

In the light of the aforesaid analysis, it is pertinent to study the socio-economic condition of Lanjia Saoras of Gajapati district (Orissa), concentrated in Gumma[1] Block in order to assess their living condition and impact of development.

Scope of the Study

The scope of the study is confined to Gajapati district of Orissa and more particularly, the Gumma block. It is one of the seven blocks of the district inhabited by Saoras, a primitive tribal group. The Saoras living in the hills are commonly referred to as Lanjia Saoras (tailed Saoras), who lead a primitive way of life. Their breather, who are living in the plains, also address them as Lanjia Saora and claim themselves superior or advanced Saoras (Jara Saora or Suddha Saora). The study is designed to explore the living condition and socio-economic life of Saoras/Lanjia Saoras—the primitive tribe inhabited in Gumma to Seranga hill areas of Gajapati district (Orissa).

Methodology

Since the present study is designed to explore the socio-economic condition of Lanjia Saoras of Gajapati district (Orissa) both primary and secondary sources of data have been collected for the purpose of present study. Gajapati district of Orissa has been selected purposively. There has been concentration of Saoras and Lanjia Saoras in this district.

Gumma block has been selected purposively for field study. As has been pointed out, there are 21 villages under the LSDA micro project with 1241 households with a total population of 5114. Hence, the study has been conducted in all the 21 villages, with the help of a structured questionnaire by personal interview method.

However, intensive study has been conducted in light villages selected on random sampling technique from among the 21 Lanjia Saora settlement villages for collection of Saora crucial information.

Data have been collected also from the office files and records of Lanjia Saora Development Agency, Seranga. The researcher has also made important personal discussions with the officials along with observations along with data collected through questionnaire and interview methods supplemented the work and finally, enriched the analysis. In fact, it was a strenuous and herculean task to collect field data from the Lanjia settlements.

Lanjia Saora and their Location

Data available on the Saoras reveals that there are a large number of Saora populations in the undivided Ganjam district, which is at the southern tip of the State. The present Gajapati district was a part of the erstwhile Ganjam district and covered almost entire hill tracts, which remained isolated and backward. The Saoras living in the hill tracts within the jurisdiction of Gumma Block of Gajapati district being referred as Lanjia Saoras. Both Lanjia Saoras and other Saoras are notified as Primitive Tribes. It is learnt from the district administration that in Gumma Block one micro project named as Lanjia Saora Development Agency (LSDA) has been established in 1979 for exclusively looking after the development of the living conditions of Saoras in 21 villages.

District Profile

Gajapati has been a separate district after carving out from the old Ganjam district in October 1992. Geographically Gajapati is bounded in the East by Ganjam district, North by Khandhamal district, and West by Rayagada district and South by Andhra Pradesh.

In Gajapati district, R. Udaygari Tehsil of Parlakhemundi subdivision, Gumma Block of Parlakhemundi Tehsil of Parlakhemundi subdivision and Rayagada Block of Parlakhemundi Tehsil of Parlakhemundi sub-division come under the tract declared as Scheduled Area.

Table 1.1: Sex-wise distribution of tribal population in Gajapati district in difficult blocks and urban areas

Sl. No.	*Blocks Urbans*	*Scheduled Tribe Population*			*Percentage of ST population to total population*
		Male	*Female*	*Total*	
1.	Gumma	18,596	20,830	39,426	71.57
2.	Kashinagar	6,771	7,031	13,802	30.09
3.	Mohana	26,007	27,072	53,079	56.87
4.	Nuagada	13,860	14,936	28,796	74.72
5.	Parlakhemundi	3,887	3,873	7,760	10.45
6.	R. Udyagiri	15,095	15,173	30,268	68.20
7.	Rayagada	21,222	22,706	43,928	77.82
8.	Urban				
9.	Kashinagar (NAC)	43	27	70	0.69
10.	Parlakhemundi (Municipality)	380	216	596	1.63

Source: Census of India, 1991.

Table 1.2: Tribal dominated districts of Orissa

Sl. No.	*Districts*	*Percentage of S.C. population*	*Percentage of ST population*
1.	Malkangiri	19.96	58.36
2.	Mayurbhanj	6.99	57.87
3.	Rayagada	14.28	56.04
4.	Nawarangpur	15.09	55.27
5.	Phulbani	18.21	51.51
6.	Sundargarh	8.78	50.74
7.	Koraput	13.41	50.67
8.	Gajapati	8.77	47.88
	Orissa	16.20	22.21

A detailed picture of tribal population of Gajapati district as per 1991 census is presented above. It is revealed from the Table 1.2 that Gajapati district is one of the backward districts (Out of 30 newly created districts) having 47.88 per cent of tribal population. The district of Malkangiri has the highest per cent of tribal population (58.36%) followed by Mayurbhanj (57.87). In the state level, there are 16.20 per cent of scheduled caste and 22.21 per cent of scheduled tribe population as revealed from the Table 1.2.

Food Gathering Habits

For gathering food also the Lanjias depend very much on the forests, apart from meeting food requirements from settled agriculture and shifting cultivation, they gather food especially fruits, roots, vegetables, meat (by poaching), honey etc. from nearby forests. Though the dependence of Lanjia Saora families having their own land is less on forests produces, the families who are either landless or are having very small sized land holding, depend largely on the forests for food gathering in almost all villages. However, the intensity of food gathering, which was being practised by all families, has decreased considerably. Apart from shifting cultivation in forest lands hill tops and slopes, some of these primitive tribe households still pursue hunting, catching small animals and birds and collecting roots and tubers. Data collected from all the project villages about normal food gathering habits of the Lanjias and availability of such food resources in the locality was analyzed.

The people were of the view that the area has gradually been depleted of its fauna as a result of which hunting has been substantially reduced. Besides, tightening of vigilance by Forest Department has restricted the quantum of hunting even though the Lanjias are non-vegetarian in food habit. From quite a good number of villages no answer was received, as the respondents felt shy to give answer on such a delicate question. Catching of birds and small animals for food still continues in summer and winter seasons in some villages even though the quantity has substantially reduced. The Lanjias and for that matte the tribals in general do not damage the fruit trees. Therefore, collection of fruits in almost all seasons is being regularly done in all villages but since fruit trees are alternate bearing a good crop is harvested by Lanjias in alternate years. Collection of roots and tubers has been discontinued in most of the villages after the project was established and greater emphasis was laid on increasing agricultural production and horticulture.

The Lanjia Saoras residing in the area have very limited wants relating mainly to food, drink and community functions. These relating mainly to food, drink and community functions. These wants are very large fulfilled from the local resources. They do not very much feel the necessity of going outside their area. Besides, they hesitate to contact outsiders or strangers are only few traders and officials move in the area for supplying some essential commodities such as salt, jiggery, dry fish and cloth etc. The clothing habit of Lanjia Saora does not require much of dressing materials. The Lanjia male uses a loin cloth about 6′ long and 10″ to 12″ in breadth which wraps around the waist and uses it from which a long portion hangs behind like a tail. Few of these Lanija males who are going outside the area are also using dhotis, as they are required to contact officials and other strangers. The females use to waistcloth, which stretches up to knees only. The skirt in only 3′ in length and 2′ in breadth. The people do not also like to go outside for medical aid or treatment of disease, as they still believe that diseases are caused due to wrath of Gods/Goddesses and evil spirits. They normally consult the traditional village medicineman (Shaman) for their treatment, which consists of sorcery, magical treatment, sacrifice of birds and beasts etc. to propitiate the Gods or evil spirits. With the introduction of project activities creation of medical facilities and distribution of medicines in the villages there has been only a marginal change and still the majority adopts traditional practices. For all these reasons, the Lanjias do not feel the need of moving out of their area besides being allergic to meet strangers.

Change in Eco-system in the Area

From the project villages some relevant information about the eco-system was collected. The primitive tribes due to their absolute dependence on the forests and other natural resource including fauna and flora were unknowingly causing gradual degradation of eco-system. Rampant shifting cultivation leading to loosen in of surface soil and large scale soil erosion coupled with poaching and killing of small animals for meat, felling of trees for getting fuel and households needs led to a continuous decline of the eco-system. Even though some restrictions have been introduced after establishment of the Micro project at Serang and some restoration has been brought about in ecology and environment, it has been possible to control the degradation fully. In spite of efforts of State Government through concerned sectoral agencies as well as through project activities, deterioration of eco-system especially in regard to forest cover, wild animals, soil erosion, land productivity and rainfall, could not be controlled substantially. Reduction of pasture lands for grazing of cattle has been reported in as many as 11 villages were as in 5 villages no change was noticed. This has seriously affected cattle grazing as the number of cattle and other domestic animals has increased which pasture lands have decreased. This has also been responsible for damage of forests by cattle. In 3 components of ecology there has been marked improvement. Incidence of diseases has been controlled effectively and provision of drinking water in project villages has improved substantially. Similarly, climatic condition and source of supply food materials have also shown the tendency of improvement. In spite of serious efforts, there has not been a significant dent in shifting cultivation even though some households having fairly large landholding sizes have reduced the quantum of shifting cultivation. From the villages it was understood that since cultivable land has been very much less in these villages located in a bad and undulating terrain, the households, depend by and large on shifting cultivation.

Drinking Water Facilities

There has been a positive improvement in drinking water supply in the project villages. This has been possible due to the efforts of the project Authorities. Before the project was established there were only 10 drinking water wells (sanitary wells) in the entire project area and the tribals were depending on water supply from nearby streams and river lets in most of the villages for drinking purposes. Due to water use from such unhealthy sources epidemics and water borne diseases were very common. Stomach disorder and skin diseases were found among Lanjia Saoras in almost all villages. This could however be controlled effectively through supply of potable water in almost all villages.

Table 1.3: Villages reporting about agencies responsible for degradation of local eco-system

Sl. No.	Components	Number of villages attributing degradation by various agencies							
		Indigent of the area	*Outsiders*	*Wild-animals*	*Natural animals*	*Natural factors*	*No idea*	*No reply*	*Total No. of villages surveyed*
1.	Forest	7	13	–	5	2	–	12	21
2.	Pasture lands	–	–	–	12	4	–	8	21
3.	Wild animals	3	11	–	–	–	7	–	21
4.	Land Productivity	8	–	–	–	10	8	3	21
5.	Climate	–	–	–	–	–	21	–	21
6.	Rainfall	–	–	–	–	–	21	–	21
7.	Food materials	6	8	3	2	–	8	–	21
8.	Incidence of disease	–	–	–	–	5	16	–	21

(Multiple answers)

It was found physically that in almost all villages some type of drinking water source is available. Since the villages are normally having very small number of households they manage well with one or 2 sources. In few big villages where the households are more scattered, some households walk a longer distance to collect drinking water from the source. It was understood from project Authorities that construction of tube wells or even sanitary wells is not possible in all villages because of high altitude and uneven surface. The water level is very low and it becomes absolutely un-economic to construct a tube well or drinking water have constructed in such villages.

Degradation of Local Eco-system

The respondents were, by and large, unaware of the eco-system as a whole even though some of them understood the various components, which constitute it. They also do not have an idea as to how the eco-system is degraded. Even after explaining the subject in detail in some cases they could not attribute the deterioration to any agency. In regard to degradation of forest even though 7 villages expressed that they are partly responsible for destruction of forests, 13 villages attributed the damages being done by outsiders who unlawfully cut down trees and remove those from the forests. 12 villages did not give any reply in this regard whereas in 5 villages the domestic animals partly damage the forests (by eating away small seedlings of forests species) as the village pasture lands are almost barren. But they admitted that they cut down no living trees. In 12 villages it was reported that pasture lands are not adequate for cattle grazing and these lands appear barren in all seasons except in rains. The small domestic animals not only graze in pasturelands but also in nearby forests sometimes damaging small seedlings. Destruction of wild animals by poaching through outsiders was very common as was reported by 11 villages even though its magnitude has, of late, been very much reduced. In only 3 villages, it was reported that some small animals and birds are trapped for collection of meat.

In spite of all out efforts to reduce shifting cultivation which is the single major factor to damage the eco-system and which is caused by the indigents of the area, there has not been any spectacular reduction in the practice even though some households having bigger parcels of owned cultivable lands have switched over to settled agriculture. The landless Lanjia Saoras as well as those having small land holdings still cling to the age-old practice to raise crops like mandia (ragi) rice, kandula, maize and oilseeds. Shifting cultivation starts with cleaning the areas after rainy seasons both by male and female members of the household while the females clear the under-growths and debris and dump those in the midfield in small heaps, the males cut down the dry trees, and creepers etc. The dry trees are cut down high above the ground level leaving fairly tall stumps. Normally the fruits trees are not touched. Soon after debris and heaps dry up fully those are burnt to ashes and the ashes are spread over the swidden. The swiddens are then ploughed if the slopes are mild and at lower altitudes but in most of the cases in cultivation is done by hand implements for sowing seeds. The swiddens are regularly used for 2 to 3 years only for raising crops and the left over, the venue being shifted to other places. Such podu cultivation normally loosens the surface soil, which is rich in plant nutrients, and during rainy season the topsoil on the swiddens is washed away. This is responsible for deterioration of eco-system in the area and decreasing productivity of land. Even though process of automatic correction continues due to deposit of vegetative deposits from the forests it takes a fairly long time to restore fertility.

Type of Houses and Materials Used

The abodes of Lanjia Saoras, the indigents of the area are very small kutchha structures with relatively higher plinth but a low thatched roof. The houses are single roomed in a majority of households with rectangular shape. It has a high verandah outside with a single entrance. The walls are constructed with stone pieces of different shapes and sizes but embedded in mud. A thick plaster of red is put on the walls to protect it and given decorated out look. A big loft is normally created inside the room with rough wooden beams resting on the walls on both sides. This loft is used mainly to store all types of household articles, food materials, utensils and even the clothing. It bigger households the size of the loft is also bigger sometimes occupying more than half of the room. The place below the loft is used as kitchen and the remaining part of the room is used for sleeping, living and even for entertaining guests. During rainy season and in winter, the members also prefer to sleep under loft. In small households, this only room also accommodates small domestic animals like goat, sheep and fowl. Agricultural and hunting equipments are either hung from the wall or fixed to the roof. The rooms are fairly dark during daytime as no windows are fixed to the walls.

The materials used for construction of houses are very simple and locally procured. The construction is done without the help of carpenter or mason and normally the family members sometimes with the help of neighbours construct their houses. The materials used are mud, stone, bamboo, grass, leaves, small timber, creepers, ropes and twigs etc. These are collected free of cost with the help of family workers from nearby forests.

Land Utilization and Agriculture

The total cultivable land in the project villages is about 570 hectares (1425 acres). This indicates that the average cultivable land per village is only 27 hectares (68 acres) and per households is only 0.7 hect. The lands are dry laying either in terrance in foot hill region or on the narrow valleys between hills. These primitive tribe households are normally engaged in shifting cultivation either wholly or partially. While about 10 per cent of the Lanjia Saora households in the project area are landless, about 56 per cent are having land less than an acre, and 22 per cent possess land with in 1 to 2.5 acres. This is the principal reason for which it has not been possible to reduce pressure on shifting cultivation substantially. The hill slopes are used for shifting cultivation.

Unless, more revenue land is reclaimed and distributed among those primitive tribal households not having land or having very small land holdings size, shifting cultivation on the hill slopes cannot perhaps be controlled. The cultivated uplands do not have any field bunds and such swiddens growing various types of crops are well visible from a distance along the hill slopes present a picturesque sight. The jhola or low lands are terraced and these are utilized for growing paddy. In the swiddens ragi, minor millots, kandula, judanga or Jhatta (bean) are raised. The cash crops grown in high lands are turmeric and ginger. The homestead land is used for growing fruit trees such as jackfruit, tamarind, mango, mauha, orange, papaya and lemon etc. and maize and chilly. Lanjia Saoras are good terrace cultivators and the terrace are valuable assets for them. Some of the terraces are having facilities of irrigation from natural springs and streams. Some terraces are also found built right up to the bed of the hill stream. While lower irrigated terraces are used to grow paddy, upper terraces grow bidi, kulthi, ragi and oilseeds.

Horticulture

Lanjia Saoras depend on fruits to a considerable extent, which they use for their consumption as well as for sale in the nearby growth centres. They do not pursue horticultural activities in a systematic way but collect fruits from the nearby forests. After introduction of the Micro project, horticultural activities have been boosted up especially in backyards of these tribal households and in the private wastelands and uplands. The intention is to reduce pressure from Government forests and to promote fruit production in the area due to its climate, topography and physical potential to grow fruit crops.

Table 1.4 depicts the village-wise distribution of lowland in the Lanjia Saora settlements.

Table 1.4: Village-wise distribution of low-level (in hectare)

Name of the S.P.	*Name of the village*	*Extent of low-land*
1	2	3
1. Bhabani	1. Bhabani	29.58
	2. Tidasing	15.35
	3. Angda	4.43
	4. Padsa horiang	3.25
	5. Abasing	10.15
	6. Arbun	6.60
	7. Gudang Gorjang	5.95
	8. Puttur	65.0
	9. Kulput	3.15
	10. Changurda	15.35

(Contd...)

1	2		3
2. Ajayaguda	11.	Bantarda	34.95
	12.	Kantisuhi	9.40
	13.	Kindurmati	1.15
	14.	Mulisahi	4.50
	15.	Radha	24.37
	16.	Luara	31.00
	17.	Telimba	18.55
	18.	Jangtar	48.70
	19.	Tipising	67.50
	20.	Adhei	24.65
3. Kajasing	21.	Raitasahi	17.40
	Total		**440.98**

Source: Office files of the LSDA, Serango (Gajapati district)

Seedlings of different types of fruits popular among tribals and profitable to yield higher income for the households have been distributed with some fertilizers and technical know how. Realizing that cereal crops grown in the area in limited cultivable lands does not meet the food requirements of these primitive tribes fully and dependence of these households on fruits with which they survive during summer months, horticultural plantation has been popularized in the area.

Jackfruit, orange, tamarind, cashew nut and mango are found in plenty in most of the project villages. Besides the homesteads and village area, the nearby forests are also full of Tamarind, Mohua, salseed, kendu, sago palm, palms and jackfruit trees. Some orchards of cashew, pineapple, orange and Banana have come up within the villages. There has not been a single village where some types of fruit trees are not available. Besides, the primitive tribes also collect a good number of minor forest produces such as brooms, Nux-vomica, Harida, Bahada, Amla, Mahua seeds, sal seeds, tamarind, Mahua flower, Lemon, Palua, gums, Jhuna etc. and sell those to LAMPS or other traders. Among the surplus agricultural produces sold by the Lanjia Saoras are jingly seeds, castor seeds, mustard, niger, horse gram and black gram etc. It was however ascertained that even though horticulture has been promoted in the area, the Lanjias have not yet been properly trained about sowing fruit seedlings and post sowing management and inter-culture etc. It is necessary that such trainings be arranged through regular training camps and demonstrations in Project villages.

Agricultural Implements and Hunting Equipments

Lanjia Saoras use animal power for agricultural operations in valley lands and terraced fields. In hill-slopes or hilltops, which are not easily approachable by big animals such as bullocks/ buffaloes, hand implements especially in steep swiddens carry on the operations. The implements used for agricultural operations were of traditional type i.e. simple manual implements such as hoe (a forked piece of wood with pointed iron tips), axe, digging sticks (pointed wooden sticks), Pickaxe, crow bar, sickle and knives. Traditional ploughs with narrow tip and pointed iron share was being used to plough the fields. However, after introduction of the project better and improved agricultural implements have been supplied to most of the farming households. Some of these implements have been quite popular among these primitive tribe households.

The hunting equipments are also of traditional type consisting of bows, arrows made of bamboo, hunting knives, spears and axe etc. For catching small animals nets and traps are also being used. Since there has been a recession in hunting activities in most of the households such equipments are no longer of any use.

Conclusion

A perusal of the afore-said analysis brings to the light that Gajapati district in Orissa is one of the tribal dominated districts with 47.88 per cent of tribal population. It is agriculturally dominated and tribals depend on horticulture and agriculture with low industrial activities. The percentage of literacy exhibits a declining trend from primary school level to college level among the tribal students with high dropout.

Notes

Gumma block is one of the oldest Community Development blocks in the state of Orissa. The block is located in the hills. Out of 212 villages, 128 villages are yet to be connected with motorable roads and 120 villages are yet to be electrified. The entire tribal population (8000 tribal households) is entitled for numerous special facilities for their development, including 100 per cent subsidy.

2

Impact of Deforestation on Environment

*Dr. S.N. Tripathy***

*Mr. Premananda Pradhan**

Forestry and agriculture are the two most important land uses in the country, the latter competing with the former under relentless pressure of an ever increasing population, which has grown from 361 million in 1951 to 1,028 million in 2001. The forest cover of the country was estimated to be 678,333 km, which is 20.64 per cent of the geographic areas of the country. Very dense forest, moderately dense forest, open forest and mangroves constitute 1.56 per cent, 10.32 per cent, 8.76 per cent and 0.14 per cent of the geographic area respectively. Scrub and non-forest areas are the other classes in the scheme of classification. The per capita availability of forests has, thus, declined to a poor 0.08 ha, which is one of the lowest in the world.

Both locally and globally, forests play a vital stand in the environmental web. Forests have a many-faceted ecological role to play which affect human life directly in a variety of ways. They thwart the dangers of cloud drifting, soil erosion, floods, wind erosion and ground water evaporation.

** Reader in Economics, Aska Science College, Aska, Ganjam, (Orissa).

* Lecturer in Economics, Science College, Konkarada, Ganjam, (Orissa).

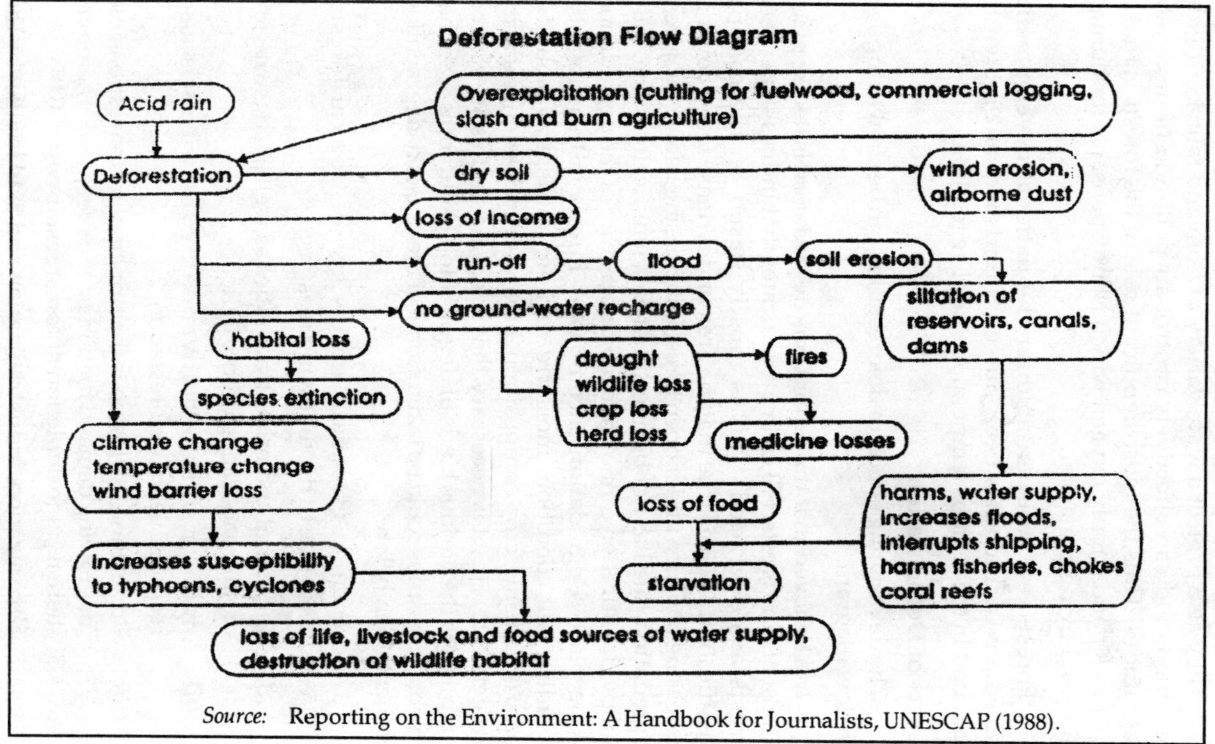

Source: Reporting on the Environment: A Handbook for Journalists, UNESCAP (1988).

They also protect a wide variety of flora and fauna, provide recreation, and can effectively control air pollution of moderate magnitudes. Thus the most significant role of the forests is that it purifies air and supply oxygen which is indispensable for human being and living organism.

Forests are indispensable for the national development. Indeed, civilization has been nursed, nourished and grown to manhood in the regions temperate natural vegetation.

Causes of Deforestation

The devastating effects of deforestation can be shown through the flow diagram.

It is believed that the continuous degradation is taking place in most of our forests. FSI estimate indicates that India's 40 per cent forests are degraded. This is mostly based on the canopy classification and it is true that the area under dense forest cover is decreasing and area under open forest is increasing. However, forest degradation rate has not been estimated scientifically, since, only canopy classification cannot give the true picture of the degradation taking place. Degradation encompasses forests in their totality, including soil condition, moisture regime, nutrient availability, micro organisms, crop composition, biomass production, capacity to regenerate, etc., it is necessary that scientific studies are carried out to assess the various kinds of degradation of our forests, so that remedial measures could be taken. Some of the causes of degradation are mentioned below.

Deforestation essentially means loss of tropical forests. Deforestation is due to the following reasons:

(i) Industrial and mining activities as well as construction of hill-road lead to deforestation;

(ii) Jhum or shifting cultivation by the tribals is also responsible for deforestation;

(iii) Population explosion is the principal factor responsible for growing demand for timber and fuel wood. Therefore, there is huge depletion of forest resources;

(iv) Urbanization, industrialization and land hunger by the non-tribals have resulted in deforestation;

(v) Without any consideration of sustainability of the forest resources, now-a-days there has been tremendous growth of forest based industries which cause deforestation.

The Devastating Effects of Deforestation on Environment

Deforestation is a critical problem. It depletes natural resources of critical importance to the people of a region, which cannot be replaced soon.

In the present day world we are in the grip of environmental problems, which was never explained earlier. In every corner of this planet, there is palpable concern about the specter of the environmental crisis. While the rich and well-off talk about Ozone layer depletion, global worming, climatic change, nuclear hazards, urban congestion and air, water and noise pollution, the poor man only understands that his crop whither in parched land, flood takes the lives of cattle and human beings and his wife known that she walks five kilometers or more daily to collect water, fuel wood and fodder.

Global warming, a consequence of environmental pollution and depleting green cover, has its adverse impact on agro eco-system as well as the bio-diversity.

The role of tropical deforestation in global warming is well-known, accounting for 20-30 per cent of annual carbon emission world wide. Felled trees emit more CO_2 as the burn or decay. When there are no trees for absorbing CO_2 automatically, CO_2 will increase in the places where trees are destroyed.

If respiration exceeded photo-synthesis for an extended period, trees would stop growing and ultimately die. It will destroy not only unique and singular plant life, but also the habitat of hundred of species of birds, insects, reptiles and mammals.

Before the advent of the industrial revolution, climatic change was a gradual process. But in the recent past, the rate of change has been triggered due to tremendous increase in pollution. Human made green house gases like carbon dioxide, methane, ozone, carbon monoxide, nitrogen oxide and hydrogen peroxide have caused the largest climatic changes in the recent years. This has ultimately affected the agro-eco-system in the following manner:

1. Evaporation will increase;
2. Water and soil temperatures will rise;
3. Soil water will become insufficient;
4. Snowing period will shorten;
5. Cultivable areas will be enlarged;
6. The incidence of insect pests, diseases and weeds becomes higher;
7. The deterioration of soil fertility will be quickened;
8. The activities of the micro-organisms will increase;
9. The decomposition of organic matter and fertilizers will be promoted;
10. Soil and shore erosion will be quickened.

A perusal of the study reveals that forestry is the imperative need for minimizing the hazards of environmental problems. Hence, an integrated approach in the management and development of foresting is required.

Conservation of Forests Through People's Participation

Social forestry programme has been expected to be a novel and unconventional measure for forestry development and management. As against the traditional form of forestry, which as an extension of the colonial legacy continued to center around commercial production and utilization of forests, social forestry aims at arresting the accelerated pace of ecological degradation, and at replenishing the loss by a foresting non-agricultural areas.

Another major difference between conventional forestry and social forestry has been that plantation, preservation and post plantation care of saplings were to be effected with the co-operation of the local people. People's participation could be evoked only by creating and sustaining their interest in plantations and only by utilizing their indigenous knowledge systems. It was expected that special emphasis be placed on involving rural dwellers to focus on the needs of women while practising social forestry.

Women's Participation and Forestry Programme

Especially rural women contribute productive labour significantly by collecting firewood, fetching drinking water often from long distances and participating in farm operations. Sowing, planting harvesting etc. It is now generally admitted that as desertification, soil degradation and deforestation process advance, responsibilities as well as sheer physical drudgeries of rural women will keep mounting.

It should, therefore, follow logically that rural women can prevent further degradation of forests and can revive the lost forest. They can rationally use and conserve forest resources in a balanced as well as sustainable manner. Women can contribute to forest conservation and management in two different ways:

(a) by economizing the use of fuelwood and cowdung as fuel (on the demand side); and

(b) by assisting a forestation of barren/denuded lands in their neighbourhood and by helping the conservation/protection efforts of forest development as well as local village communities (on the supply side).

Safe Ozone and Forest

Nature has made every possible effort to subsist mankind on earth. The best example of it is the ozone layer in the atmosphere. This thick layer of ozone protects us from the deadly Ultraviolet rays of the sun.

Man has always worked against while quenching his thirst for progress, development and new inventions. Man has created rockets to reach the upper space, refrigerators and air conditioners for enriching our lives and aeroplanes for faster mobility. Though these inventions are aimed for the betterment of the society they are in turn proving a hindrance in the way of nature disturbing the ozone layer.

A few years ago a big hole in the ozone layer was discovered over the South Pole. At that time it was as large as the size of United States. But in the recent past this hole has become so big so as to cover the whole of Antarctica and it is increasing still further. According to the scientists this hole reaches its full size in March every year and closes completely in September. Recent studies have revealed that this hole has lost its capacity of closing due to its large size.

This hole has given way to the ultraviolet rays of the sun. These rays when fall on the human body cause many diseases such as bronchitis, skin cancer etc. If these rays continue to reach at such a speed they will soon annihilate mankind.

The major cause of these holes is the CFCs (chlorofluorocarbons) which are used in air conditioners and refrigerators as a coolant. This gas is so poisonous and 'immortal' that it is not affected even by the bacteria's. It continues to rise in the atmosphere and when it comes into contact with the ozone layer it converts it into free oxygen. It is estimated that one molecule of CFC can destroy 1.00.000 molecules of ozone. So much CFC has already been spread in the atmosphere that now if we even stop its use completely then also it will take about 100 years to recover from its influence. At the present rate of CFC release, we will destroy about 18 per cent of the layer by 2050 A.D.

The great amount of heat energy is produced due to atomic explosions, and during the combustion of fuels in jet planes. This changes the nitrogen and oxygen of the atmosphere so as to combine for the formation of Nitric Oxide. This oxide of nitrogen also has the capability of converting ozone into free oxygen. Scientists have declared that 5000 mega tons of explosive creates about 50,000 mega tone of nitric oxide which can destroy 50,00,000 mega tons of ozone. The whole ozone layer is made up of 430,00,000 mega tons and only 7 or 8 such explosions are enough to destroy it completely.

The destruction of ozone layer has caught the attention of the people all over the world and a programme under the name of "Save Ozone" was held in 1989 in which 118 nations took part. It was decided that by the year 2000 A.D. all the developed countries will reduce the use of such means which are harmful to the ozone layer by about 90 per cent. The developing countries were given a concession of 10 year in this agreement.

Man is the only cause of destruction of ozone layer and man only can preserve it. Earth is the only planet which sustains life and we must make every possible effort to save our species. We can also help in this ozone saving programme by planting more and more trees.

Essential Human Needs

A critical review of the foregoing analysis brings to light the following facts. They we need the essential human needs from the forests like:

Food, fodder, fuel, fertilizer, fiberetc. The quantity of land fixed. Production needs to be increased. Scientists struggling to increase yields to match population boom.

The Critical Phase

- The depreciation in human values
- The cries of population explosion
- The concerns of environmental degradation
- The accumulation of vast quantities of waste products generated by industrial and domestic consumption—pollution
- The enormous requirements of raw materials for industry

The Population Boom

World population almost doubled during last century from 4b to 8b

In less than:

Six years is created an Europe

Three years, an America

Two years, a Japan

Malthusian Hypothesis:

"There is geometrical increase in numbers but arithmetic increase in resources".

Deforestation

- Annual disappearance rate of forests in the world–1.2 per cent
- About 11 million hectare lost each year
- This removal rate is 20 ha per minute
- About 25 million ha of deforested land exist in Thailand, Philippines and Indonesia

- Countries like Haiti have essentially eliminated all of their forests
- At current removal rates, 17 other countries will also reach total deforestation
- NRSA estimates forest cover decreased from 16.83 per cent to 13.94 per cent between 1972-83 -23 per cent; reconciled figure at 19-21 per cent
- Per annum disappearance rate 1.5 mh; forest cover can disappear in just 25 years
- For balanced environment, about 33 per cent of the area recommended to be under forests (19-21%)
- Out of this 66 per cent expected to be in the hilly areas (33-36%)

Soil Erosion

- Approximately four billion tons of soil erodes from land each year, about half of which appear as sediment in open waters. Source of this sediment are:
- 50 per cent from farm and sand lands;
- 30 per cent geologic erosion;
- 5-10 per cent from public lands;
- 10-15 per cent from roads, mine spoils urbanizing, construction activities.

On Degraded Sites

- The top soil is removed or lost, altered or buried
- The native animal or vegetation communities are lost
- Degrades sites usually become too: steep, stony, Acid compacted, infertile having low levels available plant nutrients and water storage, are unfavourable to root environment
- Natural rehabilitation of such land through normal succession process not possible
- Rehabilitation possible
- Through a new root growth medium; artificial introduction of new plant communities; altering the conditions under which new plant communities may develop

National Afforestation and Eco-development Board (NAEB)

The National Afforestation and Eco-development Board (NAEB) were set up by the Ministry in August 1992 for promoting afforestation, tree planting, ecological restoration and eco-development activities in the country. Special attention is being given to the regeneration of degraded forest areas and lands adjoining forest areas, national parks, sanctuaries and other protected areas as well as the ecologically fragile areas like the Western Himalayas, Aravallis, and Western Ghats etc.

The Objectives of the NAEB

The objectives of the NAEB are Evolve mechanisms for ecological restoration of degraded forest areas and adjoining lands through systematic planning and implementation, in a cost effective manner;

Restore through natural regeneration or appropriate intervention the forest cover in the country for ecological security and to meet the fuelwood, fodder and other needs of the rural communities;

Restore fuelwood, fodder, timber and other forest produce on the degraded forest and adjoining lands in order to meet the demands for these items;

Sponsor research and extension of research findings to disseminate new and proper technologies for the regeneration and development of degraded forest areas and adjoining lands;

Create general awareness and help foster a people's movement for promoting afforestation and eco-development with the assistance of voluntary agencies, non-government organisations, Panchayati Raj institutions and others and promote participatory and sustainable management of degraded forest areas and adjoining lands;

Coordinate and monitor the Action Plans for tree planting, ecological restoration and eco-development; and undertake all other measures necessary for promoting afforestation, tree planting, ecological restoration and eco development activities in the country.

Major Decisions Taken in the NAEB Meetings

The 11th Meeting of the National Afforestation and Eco-Development Board (NAEB) was held on 19th January, 2006, under the chairmanship of Ministry of Environment and Forests.

Major decisions taken in the meeting include:

Planning Commission may commission a study to assess adoption of joint forest management in respect of State Plan Schemes and share the same with the Ministry.

The decisions taken in the 4th meeting of National-level Steering Committee for National Afforestation Programme (NLSC) held on 14th November, 2005 were reaffirmed by the Board. The major decisions of the NLSC included:

(i) initiating steps for strengthening the monitoring mechanism of NAP by activation of State-level Coordination Committees for NAP, increased field visit by officers and active involvement of Divisional Commissioner and District Magistrates in the process to review the increase in forest and tree cover;

(ii) Strengthening the support staff in the office of Conservator of Forests and Divisional Forest Officers for the account of Forest Development.

NAEB operates the following three major schemes:

National Afforestation Programme (NAP), Grants in Aid for Greening India (GIA for GI)

Monitoring and Evaluation (M & E) Communication, Support to Regional Centres (RCs), Eco Development Forces (EDF) Scheme, Programme-wise progress of Activities of NAEB is as follows:

National Afforestation Programme (NAP)

It continues to be the flagship scheme of NAEB, is so much as it provides support, both in physical and capacity building terms, to the Forest Development Agencies (FDAs) which in turn are the main organ to move forwards institutionalization of implement Joint Forest Management. The FDA has been conceived and installed as a federation of Joint Forest Management Committees (JFMCs) at the Forest Division level to undertake holistic development in the forestry sector with people's participation. This is a paradigm shift from the earlier afforestation programmes wherein funds where routed through the State Governments. This decentralized two-tier institutional structure (FDA and JFMC) allows greater participation of the community, both in planning and implementation, to improve forests and livelihood of the people living in and around forest areas. The village is reckoned as a unit of planning and implementation and all activities under the programme conceptualized at the village level. The two-tier approach, apart from building capacities at the grassroots level significantly empowers the local people to participate in the decision making process. Under Entry Point Activities, community assets are created with a 'care and share' concept. The objectives of the scheme are given below:

Protection and conservation of natural resources through active involvement of the people, checking land degradation, deforestation and loss of biodiversity, Ecological restoration and environmental conservation and eco-development, evolving village level people's organisation which can manage the natural resources in and around villages in a sustainable manner, fulfillment of the broader objectives of productivity, equity, and sustainability for the general good of the people. Improve quality of life and self sustenance aspect of people living in and around forest areas. Capability endowment and skill enhancement for improving employability of the rural people.

The NAP Scheme is being implemented through two-tier decentralized mechanism of Forest Development Agency (FDA) at Forest Division Level and Joint Forest Management Committees (JFMCs) at the village level.

Grants in Aid for Greening India

The earlier Grants-in-Aid to Voluntary Agencies Scheme have been restructured as "Grants-in-Aid for Greening India" during the year. It has two subcomponents:

(a) Production of, and awareness generation about high quality planting material; and (b) Grants to voluntary agencies and other bodies for taking up plantation of high quality planting material. For production of high quality planting material (QPM), support is provided for setting up of Central nursery by the State Forest Department and satellite nurseries by various organisations and individuals. Provision for certification of the (QPM) seedlings has also been made. In addition, the state forest departments are also supported in awareness generation about use of quality planting material through different publicity measures like posters, pamphlets, etc.

Planting of only high quality seedlings is to be taken up under the scheme.

Eco-Development Forces (EDF)

Eco-development force was established in 1980s as a scheme being implemented through Ministry of Defence for ecological restoration of terrains, rendered difficult either due to severe degradation or remote location or any other exigencies. The scheme of Eco-development force is based on twin objectives of ecological restoration, and employment generation for able bodied ex-servicemen.

Under this scheme, the establishment and operational expenditure on the Eco Task Force (ETF), Battalions raised by Ministry of Defence (MoD) is reimbursed by Ministry while the inputs like sapling, fencing, etc. as also the professional and managerial guidance is provided by the State Forest Departments.

In Eco Task force battalions, the MOD deploys its ex-servicemen, preferably from within the area of operation, whereas the nuclear core of the force is constituted of regular servicemen. Some of the ETF, Battalions have undertaken successful eco-restoration of highly degraded sites, for example the limestone mining areas in the Mussoorie Hills.

Joint Forest Management Cell (JFM)

Joint Forest Management (JFM) cell has been established in 1998 to monitor the JFM programme and generate policy responses. Major activities of JFM Cell are as follows:

Guidelines which envisages involvement of women in JFM activities, extension of JFM in good forest area for better NTFP management, legal back up to Committees, signing of MOU between states and JFMCs to give them sense of belonging and integration of working plans and micro plans have been issued.

JFM Network for consultation between stakeholders and for policy development has been operationalised.

Conclusion

In the context of the aforesaid illustrations, it is obvious that the physical and biological interdependence between plants and animals is fundamental and natural truth.

They can never be mutually exclusive for sustaining their existence. Forests not only produce nutrition and live air to biological world but also indispensable for preserving ecological balance and also for stability in bio-systems.

Thus, it is pertinent to suggest that for soil, water the ecological conservation, three should be forestry development. In addition for the disaster mitigations such as flood and drought, plantation of trees and development of forests are indispensable.

REFERENCES

1. Om Prakash Meena, Amit Singh Charak, B.L. Gour, *Global Warming and Its Impact on Agriculture*. Kurukshetra, June 2005.
2. Prof. Gupta, V.S. *Deforestation and Its Consequences*, Kurukshetra, March 2004.
3. Dhanagare, D.N. *Joint Forest Management People, Panchayats and Women*. EPW Sept. 9, 2000.
4. Dr. Pandey, D.K. and Tripathy, Ram Ashery, *'Role of Forest on Environment*, Yojana, April 1998, Vol. 42, No. 4.
5. Tripathy, S.N. and Pandas, *'Fundaumentals of Environmental Studies'*, Vrinda Publications (P) Ltd., 2nd Revised Edition.

3

Education of Children with Special Needs

*Dr. S.N. Tripathy**

An attempt has been made in this paper to focus on the rationale of education of children with special needs as well as exploring the prospects of strengthening such education on the basis of ground reality.

Because of the close association of the researcher with the village level non-profitable charitable organization known as SAIMA, in the state of Orissa on mentally retarded children, observation method has been adopted to infer findings of the paper. The analysis portrays the paradigm shift from a charity-based approach of educating the physically challenged to a right-based approach and ultimately to move towards the human rights perspective.

Introduction

A small segment of our child population constitutes disabled or physically challenged whose development is not only a matter of human resource development but also a humanistic imperative. It is both an economic necessity as well as developmental needs ensuring social justice.

*** Principal Investigator, the Major Research Project and Reader in Economics, Aska Science College, Aska (Orissa) 761 111.**

There has been paucity of data and no systematic survey with regard to mentally retarded children in our country. The statistics of National Sample Survey (1991) reveals that the population of visually impaired is 4.005 millions; hearing impaired 3.242 millions, speech impaired 1.966 millions, hearing and/or speech impaired 4.482 millions; and locomotors impaired 8.939 millions. In general all children must be given the opportunity to learn, whether they are average, bright, dull, retarded, blind, deaf, crippled, delinquent, emotionally disturbed, or otherwise limited or deviant in their capacities to learn.

It has been estimated that about 15,000 schools have enrolled about 60,000 children under the Integrated Education of Disabled Children (IEDC) scheme of the government of India (Singh, 2003) According to 'National Commission on Teachers', "not more than 5 per cent of the blind and deaf children and perhaps 0.50 per cent of the mentally challenged children are estimated to be enrolled in about 800 to 1000 special schools of our country. Most of the schools are located in metropolitan cities and other urban centres, whereas rural areas where about 80 per cent of these children are located remain practically unnerved by special educational facilities." The Government and Voluntary Organizations and more particularly the advent of SSA have exhibited intense and increased interest on attending to the strategy of 'how to reach the out-reached population of disabled.

Special children are differentiated as intellectually high or low; deficiency in ability or efficiency in ability. Children pertaining to be necessitated with specially designed methodology of learning who are intellectually impaired, physically impaired or both are considered to be special children. Special child has been defined as a child who deviates from the average or normal child in mental, physical or social characteristics to such an extent that he requires a modification of school practices or special education services in order to develop his optimum capacity (Manivannam, 2001). His needs must be attended to through special education which has been defined by UNESCO (1968) as "an enriched from of general education aimed at enhancing the quality of the lives of those who labour under a variety of handicapping conditions and thus, bears testimony to the concept of equality of educational opportunity".

Special education is not a total programme which is entirely different from the education of the ordinary child. It refers only to those aspects of education which are unique and / or in addition to the regular programme for all children. For example, the general educational programme for a child with visual loss is carried out in all phases by his regular classroom teacher. The special part of his education such as learning of Braille, abacus, orientation and mobility, etc., is dealt by the special teacher during the leisure time of the child. To understand the needs and for the spreading of special education, University Grants Commission (UGC), as early as in 1984-85, encouraged the special education activities in the University departments and Colleges of Education to start special education teacher preparation centres. However, currently, in India.

Inclusive Education

Inclusive education is much more than children with disabilities being allowed to go school. It includes a range of strategies and changes with in a community or society which will ensure that 'all' children have access to education.

Special education has been based on the assumption that there are children who have special educational needs. In the special education system the children with special education needs require special teachers, special resources and special methodologies of teaching. Special schools were developed when other schools would not accept students with disabilities.

Inclusion Means

All students are integrated into regular class rooms and activities for both education and social opportunities, receive on individualized education programme with support enabling them to learn in the regular class room, attend schools in their neighbourhoods and do not attend segregated programmes any where. It means children with special needs should be placed in the regular school classroom and therefore, there should be no condition imposed, to exclude him/her from that environment.

There has been continuing debate on the issue of effectiveness of separate and inclusive models, but the recent trend is towards inclusion. This right has been upheld in the universal declaration of Human Rights (1948), the convention on the rights of the child

(1989), the world declaration of Education for All (2000), and the Millennium goals (2000). The convention on the Rights of the child mandates that states makes primary education compulsory and available free to all children, on the basis of equal opportunity, with protection from all kinds of discrimination, including discrimination on the basis of disability. It also requires that children with disabilities have assess to and receive education in a manner which will help each child to achieve the fullest possibilities of social and individual development.

To optimize student performance a teacher usually works largely towards effective teaching and managing class room behaviour. But there is certain non-class room setting that a teacher needs to pay attention to. Students with mental retardation are exposed to wide spectrum of opportunities of learning in regular classroom and non-regular classroom settings like playground, school premises etc. The teacher can facilitate social learning, react to problem as and when they emerge so that the students with mental retardation are though not physically but socially integrated. The following methods can be applied in order to maximize the learning in the classroom:

1. Build motivation to learn in all possible ways
2. Use a system of rewards for appropriate behaviour
3. Relate learning situation to the areas of student's interest
4. Pre-requisite skills, information and concepts must be convenienced to the students before proceeding to new course or tasks
5. Using concrete examples whereever necessary
6. Maximum use of group experience is necessary
7. Opportunities for verbal expression should be created

Sarva Shiksha Abhiyan (SSA)

At the time of Independence in 1947, the literacy rate was below 25 per cent of the total population 350 millions. Over the years even though progressive development is visible in literacy rate, the education for disabled has received inadequate attention. Education changes the perspective of human life in all dimensional such as social, cultural, environmental and genetically. The aim of education is to provide opportunities to all children irrespective of caste, creed, colour and disability as focused in the concept of "Education for All" (EFA). Article 45 of the Constitution of India recognizes education as a basis right of every child.

The development of child's personality and potential is obstructed by chronic poverty, consistent illness and mental disorder. A child with physical or mental impairment or chronic illness has to confront numerous difficulties in carrying out its day to day activities and communicating with others. Such disabilities are to tackled through the integrated approach of education and health development or rehabilitation in the special schools or through inclusive education otherwise; these problems will be multiplied under adverse social circumstances.

The government has launched the SSA, the National Programme for providing quality of Elementary Education to all children including children with special needs in partnership with the states for the children in the 6-14 age groups through active participation of all functionaries. It covers 9.72 lakh existing primary and upper primary schools and 36-95 lakh teachers (Economic survey, 2005-2006, Government of India). SSA framework has made adequate provisions and ensures that every child with special needs irrespective of kind, category and degree of disability, be provided meaningful and quality education. Apart from this, the SSA also focused on free textbooks for girls, toilets especially for girls, setting up of Village Education Committees.

Education to the Mentally Challenged Children Through SAIMA

Social Aid Improvement and Mass Action (SAIMA), a pioneering non-profitable charitable organization has been established and registered under society registration Act 1860 since July, 1995, in a rural region of Kodala in Ganjam district of Orissa. The organisation is the outcome of the dedicated and committed educationist, teachers and local community including the researcher of this study. Out of its multi-dimensional activities, welfare and education of mentally retarded children is of paramount significance.

The research findings of the study reveals that about 2 per cent of the poverty stricken population who mostly reside in rural regions have been victimized by mentally challenged children, out of which $3/4^{th}$ are in the category of mild mental retardation and $1/4^{th}$ are severe mental retardation.

Realizing the gravity and seriousness of the problems, SAIMA has undertaken special education and training programme on behaviour modification and group activities known as VALVIKAS for the well-being of the downtrodden and weaker section of the society. The organisation has done commendable work in the realm of socio-economic survey and identification mentally challenged children and their admission in to the special school with the help of 3 bridge teacher and one special teacher. The teachers were provided training at the Secunderabad (Andhra Pradesh) to deal with the psychological problems of the mentally retarded children of the remotely situated school to benefit the rural children of Ganjam district (Orissa).

Children below the age of 3 has been admitted to the early intervention programme while those between 3-15 years have been enrolled in the VALVIKAS School of education programme. As most of the mentally challenged children are beset with the problems of speech problems, language defects, the school must have the provision of taking care of such children putting emphasis on individual needs, care and consideration. The specific problems and behavioural abnormalities of each child have been studied in depth in order to provide special attention to their needs.

The mental retarded children of VALVIKAS have exhibited their competence and confidence by winning numerous awards and accolades in special contest for the mental challenged held on the occasion of annul day celebration of the school or even at the district and state level in painting, fancy dress, music, doll making, group dance etc. The task of the future is to identify the ways in which the schools, as part of social environment can create better learning opportunities for all children and by this means, address the challenge of learning difficulties in the school system as we conceive that practical implementation of the strategy of educating the children of special needs is more important.

SAIMA also provided vocational training of preparing office files, greeting cards, weaving door mats, screen printing, tailoring and electrical coil winding etc. to help the mentally challenged children in order to integrate them in the mainstream of community life. It has been observed by the researcher that about 70 per cent of the mentally challenged children have speech defects; individual attention has to be given to each child by a qualified speech therapist and other trained staff. Students with mental retardation sometimes make efforts to learn on their own. But their effort may not be successful each and every time, and may lead to failures or disappointments. Thus, the child may be disheartened and will not be interested to learn further. Therefore, the teacher may use rewards skillfully and effectively while training or teaching the students with mental retardation.

It is a misconception that children with mental retardation can not become the part of the inclusive education system. Children with mild mental retardation will cope in regular education in primary and upper primary schools. The package for inclusive education at pre-school level has been developed by the National Institute of Mentally Handicapped (NIMH) with funding support from UNICEF. It provides a tailor made curriculum and instruction for regular education with adoption in one teaching method for the content for each disability at pre-school level.

Children with mental retardation have considerably impaired intellectual functioning and adaptive skills like self-care, communication, social skills, community care, leisure and work. Teaching consideration for children with mental retardation calls for appropriate methods or specific strategies in terms of environment, teaching, flexible and functional curriculum and aids. The benefits of inclusive education will be both to the normal and mentally challenged children.

Suggestions

Major stumbling blocks to the provision of quality education for children with disabilities in all educational contexts, include the lack of early identification and intervention services, negative attitudes, exclusionary policies and practices, inadequate teacher training of all regular teachers to teach children with diverse abilities, inflexible curriculum and assessment procedures, inadequate specialist support stuff to assist teachers of special and regular classes, lack of appropriate teaching equipments and devices, and failure to make modifications to the school environment, to make it fully accessible. These huddles and impediments can be overcome through policy, planning implementation of strategies and allocation of resources to include children with disabilities in all national health and educational development initiatives available to non-disabled children.

Because of their very special needs children are the most vulnerable, constantly exposed to risks of ill-health and disability. Caught in the vicious circle of malnutrition and infection, several million children die or else do not grow to their full physical and mental potential. Hence, we should conceive the needs of such children and provide them their physiological and psychological requirements for their development mind and body. Maternal health, nutrition, immunization, controlling communicable diseases and provision of education have significant contribution in improving their mental and physical health.

Lack of access to education is the dominant factor reinforced due to poverty leading to exclusion of all children, both, those with disabilities, and those without. Exclusion from education for children and youth with disabilities, results in exclusion from opportunities for further personal development, particularly reducing their access to vocational training, employment and income. It limits their active participation either in the family or in the community (Penny Price, 2003) Failure to access education and training can not make them independence in economic and social sphere and enhances the vulnerability to life-long miseries with poverty and pessimism.

A committee should be made up of a psychiatrist, a paediatrician and a speech therapist under the district co-coordinator of SSA in each district in order to visit the special schools or even the schools with inclusive education of the whole district at least twice in a month or at regular intervals, for treatment of such children.

In order to strengthen the mind and body, all the students under inclusive education must practice yoga. This not only improves the adaptive behaviour but also general health of children.

They are also to be encouraged to participate in cultural, recreational, games and sports which will bring about all round personality development.

Education policy and planning should be evolved so as to accommodate children with disabilities in the primary schools. Families and organizations of people with disabilities are to be consulted and involved in the process of education through the mission of SSA.

REFERENCES

1. Manivannam, M (2001) *Educating Special Children*, Social Welfare, New Delhi, March, pp. 10-11.
2. Penny Price (2003) *"Education for All (EFA): An Elusive Goal for Children with Disabilities in Developing Countries in the Asia and Pacific Region"*, Asia Pacific Disability Rehabilitation Journal, Vol. 14, No. 1, pp. 3-9.
3. Singh, J.P (2003) *Disability Status of India*, Rehabilitation Council of India, New Delhi, pp. xxiii.

4

Empowerment of Women and the System of Panchayati Raj in India

*Sobhan Kumar Patnaik**

Abstract

The local governance happens to be a vital component of democratic decentralization postulated in the Directive Principle of State Policy of the Indian constitution. Mahatma Gandhi, the Father of Nation is an unrivalled exponent of Panchayat Raj System where everyone has a share in the administration. Lord Bryce, a harbinger of democracy expouses that the democracy in a government in which every one has a share.

Mahatma Gandhi had aptly stated that a country cannot prosper unless and until the women take part in public life. Decades back he had stated that democratic process and empowerment of women should go hand in hand. The 73rd Constitution Amendment is a shot in the arm in the system of Panchayati Raj and its corner stone being power to the people. It has created a space for the political empowerment of weaker sections and active participation of women representatives in the fabric of PRI. The concept of power to the people cannot fructify unless the women who constitute 47 per cent of Indian population are empowered for it.

* **Research Scholar, Women's Studies Research Centre, Berhampur, Orissa. E-mail: sobhan21@rediffmail.com, Phone: 09937800502, 0680-2242377, Fax: 0680-2243764.**

The 73rd Constitutional Amendment is a landmark transition of political power to the women with the provision of 33 per cent of seats to women. The indication of women in large numbers is a great striding step for the rightful representation in the operation of power and political empowerment of women. However, the Panchayati Raj system in our country provides good opportunities for participation of women. But some impediments stand in the way. The maladies stemming from socio-economic, cultural and political exploitation are to be encountered and tided over. The sufficient women power to be generated by organised efforts. To effect organisation for working together in groups, to understand the problems of the society, to find suitable solutions and enlighten them the constitutional provision and to create awareness in this domain are the imperative necessity. Another important dimension should be the involvement of women in the decision making process so as to encourage the women folk to acquaint themselves with the realities and arm them with the prudence and potential to counteract the consequential problems. It is only women empowerment which can propel the women to take over the responsibilities, to comprehend the process and procedures of local governance.

Introduction

In fact, women like the poor, have been conditioned to a culture of passive silence, and they need to be put through a process of enabling of empowerment, they could articulate adequately and effectively. And as such, they need to be focus of special attention—and even exclusive measures, until they could be counted upon to withstand the silencing forces. Women's power in man's world doesn't seem to get the attention it deserves. We do seem to need a new slogan: Women's right in Women's hands". When women have space and time to meet, their pains get shared, their concerns get articulated, their goals get classified and their energies get galvanized. And they become a power. It is empowerment that comes from within. It is recognised that the initiative and efforts taken to address women's issues have been adequate, distorted, and vertical top-down and have rarely emerged out of women's priority concerns.

In India, the institution of local governments has been almost defunct since many decades. It became essential to revive them because of the realisation that unless the administration at the lowest level became democratic in its functioning, it would not be possible to create a true democratically administered country. Moreover, the failure of centralized planning and the success of the efforts at a decentralized system, at the village level, at the initiative of communists in West Bengal which attracted nation wide attention, have undoubtedly contributed to a rethinking on the need for devolving power to the village. It also became necessary to help develop proper leadership at the village-level by making it democratic in its functioning by ensuring a proportionate representation for different sections of society, including women. This paper discusses women's participation in Local Government and the need for their empowerment.

Women constitute the only "group" which has been treated unequally as a whole in spite of their comprising the largest "group" in the world as a majority of humankind. There has been an increasing need that this group gets into democratic processes and institutions. Because, the democratic system that assures equal citizenship and rights to women all over the world that believes even the poorest of the poor have a basic human right to take part in making the decisions that affect their lives and livelihoods, is proving itself to be in no position to guarantee or implement these assurances or put its beliefs into realization. Women gaining entry into the functioning of state and sharing control over the apparatus of the state alone would lead to preserve and promote equality and justice and bring about the necessary policy changes and social restructuring.

Women are endowed with reason and are capable of making rational decisions. The notion that women are "Naturally" indifferent to politics is a mere prejudice. There must be something that prevents their involvement. It may be as a result of generation of neglect the real capacities and skills in women have been stunted. Then justice demands to eliminate or moderate obstacles that are found in the way of women's participation.

Since women have materially different position in society, they have objectively different interests from men. The sexual differentiation in conditions and experiences has produced specifically women's point of view, which is either complementary or antagonistic to the men's. Different experiences do create different values, priorities and interests. Where there are different interests and different experiences, it is either naïve or dishonest to say that one group can speak for the entire society and that women had no need for a separate voice.

Studies on women have explored whether women representatives do indeed act differently from men, whether they adopt a different style of political behaviour, identify a different range of priorities, are more liberal or radical than men. The implication is that they ought to be different, and if they are not, it undermines the case.

Sometimes the arguments are made and it is argued that women would bring to politics a different set of values, experiences and expertise. The content of politics will change if women get into politics. They would enrich the political life, usually in the direction of more caring and compassionate society. When women act politically, they often do so in ways that are different from those of men. Movements in which women predominate, tend to be less hierarchical, to be more likely to be both progressive (i.e. seeking a better life) and offensive (i.e. challenging the status quo) than movements in which men predominate. (Frank and Fuentes, 1987).

Women compose one-half of the world's population and perform two-thirds of the world's work hours, yet are everywhere poorer in resources and poorly represented in positions of decision-making power. The secondary status of women in society is one of the true universal, a pan-cultural fact. Yet within that universal fact, the specific cultural consumptions and symbolizations of women are extraordinarily diverse and even mutually, contradictory. Further, the actual treatment of women and their relative power and contribution vary enormously from culture to culture, and over different periods in history of particular cultural traditions.

In India, the social structures of inequality such as caste, community, class and geographic setting create different problems for women of different spheres and these structures are critical determinants of women's troubled status. Some aspects of every religious community's law discriminate against women and these laws are still recognized by the courts. Social constrictions prevent women from significant participation in the economy and its politics.

In the first ever gender gap study covering 58 nations, the World Economic Forum has ranked India at a low 53rd position. The report measures the gap between women and men in five critical areas like economic participation, economic opportunities, political empowerment, access to education and access to reproductive health care. One of the most basic indicators of Gender inequality is the distribution of power between men and women. In India, women in general and rural women in particular continue to constitute the single largest group of backward citizens who neither have access to power structure nor are able to overcome their age-old inequality and subordination. Various studies refer to the fact that marginalization of women in all sectors, including political one, despite gender equity in various legal and constitutional provisions.

There are ideological constraints as opposed to constitutional constraints that reinforced the traditional laws, the man should govern the world outside the household and women should follow his leadership. Women are strongly influenced by the false motion of their backwardness and inferiority and the belief that social and political activity is none of their business which had cultivated in the course of many centuries. In addition, the institution of Purdha, severely restricts female participation in leadership.

Women acquire genuine equality only with the abolition of all forms of oppression. Women encounter with difficulties in obtaining their husbands' consent to political participation. The costs of electioneering and the pressure of domestic duties, coupled with the traditional dominance of men in politics, are all seen as obstacles to the participation in politics.

Concerned with the progress of rural women, economists argue that "unless there is significant change in the planning strategies and instruments of development and in the attitudes accompanying the increased participation of women in social production, the extent of women's progress will be limited."

As regards the political status, it has been observed that, "unless the economic and social utility of women enhanced in the eyes of their families and the nation by opportunities to take part in economically and socially productive roles, the national neglect of women will continue. The Ashok Mehta Committee" also noted that "Non recognition of women's role and contribute to the rural economy due to the prevalence of middle-class bias in the development planning as a social invisibility of these women due to their under representation in strategic positions in local and middle level decisions making institutions."

To bring equality to women is to give them power on equal chance with men, so that they can play an effective role in the decision making process, especially in areas where various policies have direct bearing on their well-being. Participation of women in Panchayati Raj institutions has been considered as the most effective instrument to remove their inequality, invisibility and powerlessness.

Prospects of Women Empowerment Through PRI

The historic transition of political power to women through Constitution's 73rd Amendment, with the provision of 33 per cent seats to women is a significant landmark on the road to women's equality. Also induction of women in great numbers to enter political stream, as a matter of constitutional right, is the first major step towards achieving the objective that the women should have their rightful representation in the operations of power and would be a major move towards the political empowerment of women.

Panchayati Raj, though provides opportunities for the participation of women, serious difficulties stand in their way, struggle for equality and for participation, as equal partners at all levels of decision-making. Progress towards equality will mean eliminating some of the injustice of the prevailing socio-economic, culture and political exploitation of women and making women aware of such exploitation. Organised effort, therefore, is needed to generate sufficient women power in the rural areas to fulfill the statutory representation, as per the reservation; to enable them to recognise that many of their problems are the result of social condition rather than their personal defects; to recognise the necessity for working together in groups to affect social change; to understand the problems of society, so as to find suitable solutions for those problems to make them aware of the Constitutional. Amendment and the new Panchayati Raj system; and to educate them to exercise their power judiciously.

Contribution of women to democratic development should be meaningful. At present, a large number of women have little experience of public activity and many are politically immature. Women should be educated first, as democracy will not take firm root and will have only stunted growth in a society where ignorance and illiteracy reign supreme. Statutory representation of women includes their involvement in decision-making process about what would be done and how; their involvement in drawing up schemes of productive employment generation, in utilizing local resources effectively and in implementing programmes and or their involvement in efforts to evaluate such programmes. The political participation would also amount to their sharing in the benefits of development programmes.

The soul of democracy is a medium for transmitting a sense of participation in the process of evolution of society for the governance and development of the country. Participation as a process of empowerment and self realization is valued as both right and responsibility.

The enactment of Seventy third Constitutional Amendment Act with a view to involve women in decision-making provides psychological empowerment and a sense of political efficacy to those who have felt powerless to influence public decision that affect them. With the parameters of Local Government Activity, women learn new skills, gain self-confidence, achieve greater autonomy as individuals and feel a sense of unity with the members of the concerned Panchayati Raj Institution.

Panchayat elections has brought various contributory factors to light. There was general assumption that Panchayat women being semi-literate or illiterate and therefore irrational can neither contribute to their own development nor to the development of the society. Today, it is very difficult to sustain such assumption for there are many number of examples to show how elected women, though participatory approach have been managing their local problems and to overcome them very effectively.

Elected women in some states have taken part in protest and rallies, against low water and for drinking water. Some reported to have participated in discussions for better implementation of Government scheme and communicated information regarding schemes of Government to the people.

It is becoming clear that when women see the possibility or the need to improve their situation, they become 'visible' and achieve, even to the point of taking up militant leadership roles. In Arunachal Pradesh, some of the newly elected Panchayat Women leaders even faced imprisonment in the process of persuading the government to build 12 bed hospital for women.

Throughout history women-individually, collectively, and sometimes with men have struggled against direct and indirect barriers to their self development and their full social, political and economic participation. Women need power to be able to change things. Power is inherently linked with empowerment. "Empowerment" a concept defined by Morgen and Bookman (1988: 4) as "a spectrum of political activity ranging from acts of individual resistance to mass political mobilizations that challenges the basic power relation in our societal, activities that are carried on in the daily lives of ordinary people and tare embedded in the social institutions and political economic processes of their society.

Empowerment of women requires the creation of the atmosphere which both encourages them to feel that they can effectively participate in local governments and helps them to acquire the confidence to make a full contribution with the best of their skills in decision-making.

Individuals form the basis of any empowerment. The first step in making an empowerment initiative take hold at an individual level is to gain women's commitment to the idea of change in general. Individuals become supportive of the committed to change, when they understand that the change involves and why it is important; when they understand and accept implications for themselves and the community; when they believe they will be supported in bringing out the change.

To bring out behavioural change, women should be made aware of the need for change at a personal level. Individuals can only make genuine changes if they understand the underlying purpose and reason. Real change comes from internalizing the problem and accepting responsibility for doing something about it. There should also be acceptance to the need for behavioural change.

Many empowerment initiatives fall to unlock the potentials of individuals despite extensive training because of fierce resistance to empowerment. Overcoming the barriers to empowerment within individuals requires coherent approach to: transforming resistance into positive change; ensuring that Panchayat women understand and commit to the purpose behind the empowerment initiative; building initial skills and confidence; gaining commitment to changes; building initial skills and confidence; gaining commitment to changes; and supporting people with information they need to make use of their new-found skills.

Women can only achieve visibility through empowerment. Being empowered they would be able to take over responsibilities, understand the process and procedures of local governance and with enlightened minds, they would do the right thing in the right manner at the right time and at the right place.

PRI and Tribal Women in Orissa

Tribes constitute about 23 per cent of the total population in Orissa and tribal women constitute slightly more than 50 per cent of the tribal male population. They are economically, socially and politically backward and languishing in poverty. The percentage of literacy is below than 19 per cent among the tribal women who mostly reside in forest and interior pocket of Koraput, Keonjhar, Sundegarh, Kondhmal districts of Orissa. In the above backdrop let us focus on PRI and possible empowerment of tribal women. The major impetus, in recent years, in bringing about a change in the attitude of women has come from the freedom movement. But the status of tribal women in Indian culture has been relegated in to abysmal low. In era of gender justice, reservation of seats for women and their participation in PRI process is objectified to imbibe the spirit of self-governance, self-reliance, empowerment, concretization, mobilization and organisation of grassroot disadvantage group to eliminate exploitation, injustice, poverty, social inequality etc. (Sahoo, 2000). In contrast to these basic propositions, majority of women members were elected uncontested and occupied positions in PRI unwillingly under force and compulsion of local non-tribal elites whose associates could not contest in election because of reservation. None of them were aware about role and responsibility and governance system of PRI. They neither attend any periodic meeting of Palli Sabha, Grbha, Sabha and related meeting of Panchayat, block and district nor participate in any decision making process. In this regard, the objective of bringing women-folk to PRI through reservation system for effective sustainable development of grassroots marginalized section marked obsolete and obfuscate. So, far these elected women representatives wield only formal political power without any scope of reinforcing or alerting of existing power structure and relations at the local level. Hope, the struggle of tribals in general and women in particular through PRI governance system and political will of the state could ease the problems of their life, otherwise the very purpose of decentralisation system will be in a deadlock.

Suggestions

1. Political participation requires information, knowledge and exposure to the various experiments strategic and action which coupled with training and constant interacting. This very idea should be given importance.
2. Some special institution like Mahila Mandals and Samajas should be created to promote, develop and maintain women's participation.
3. Continuous efforts should be made to train the leadership with a broader perspective than before. Comprehensive and meaningful training programmes have to be designed for rural women and they can perform their duties and responsibilities as elected members of the PRIs.
4. The reservation for women is a progressive effort. But the present Women Bill fate has been jinxed and become an insidious attempt. In fact, women's demand for reservation will be a greater respect and enhance their right in a wider democratic movement.
5. Every state should have set up a Women's Development Corporation to promote, enlightenment and employment schemes for women belonging to the weaker section of the society.
6. Many women are not aware of their right. Thus the move to enforce laws must be accompanied by a movement to

bring about an attitudinal change in our society. A Commissioner for Women's Rights to follow and monitor enforcement of legislation in this regard seems to be necessary.

7. National Resource Centre should be set up which can act as an apex centre for gender issues, including efforts of sensitization training, evaluation, documentation, information, dissemination and networking.

8. The women need to be trained in the art of administration, polity, finance, and the cybernetics of development.

REFERENCES

1. Bookman, Ann and Sandra Morgen, (1988), *Women and the Politics of Empowerment*. Philadelphia, Temple University Press.
2. Dilys M.H., (1974), *Democratic Theory and Local Government*, The New Local Government Series No. 12, London, George Allen and Unwin Press, p. 20.
3. Frank A.G., and Fuentes, M'Nine (1987) *Theses on Social Movements, IFDA Dossier* 63 (January/February): p. 27-44.
4. Kanango S.K., (1996) *Panchayati Raj and Emerging Women Leadership: An Overview*, Social Action, Vol. 46, No. 1, January-March.
5. Phillips, A., (1994) *The Representation of Women*, in Polity Press (ed). The Policy Reader in Gender Studies.
6. Sekhar, Helen R., (1994) *Women's Participation in Panchayati Raj*, National Labour Institute News Letter Vol. 2, No. 5, October.
7. Sekhar, Helan R. (1996), *Towards Making Success of Their Political Equity: Women in Panchayati Raj*, p. 12-15, February 1996.
8. Sahoo, U.C., (2000), *Panchayati Raj and Tribal Women Members*, in Tripathy, S.N. (ed) in Glimpses on Tribal Development, Discovery Publishing House, New Delhi, pp. 65-66.

5

Indian Water Resources, Aqua Food Production and the Prospectives of Silver Revolution in India

*Dr. Tanuj K. Bisoyi**

The Component of Indian Water Resources

India possesses a vast water resource both fresh water and marine water. When the fresh water resources includes the water resources zones like rivers, canal, reservoirs, tanks, ponds beels, oxbows, and derelict water etc. and the marine water resource includes the a water resources of costal lying or marine water resources. We also include brackish water area as our another source of water resources.

Aqua Food Production

There are two types aqua food production such as (1) the Inland fishers of fresh water recourses aqua culture; (2) marine water fishers or sea water resources aqua culture. The Inland fisher recourse comprise of rivers, canals, estuaries, floodplains wetland, lagoons and reservoirs and this sector holds enormous production potential to meet the inland fisher requirement of the country. The rivers system of country comprises 14 major rivers (casement>20,000 Km^2), 44 medium rivers (casements 2000 to 20,000 km^2) and innumerable small rivers and desert streams different rivers system

* **Dr. Bisoyi, Senior Faculty, Department of Economics, Kabi Surya Baladev Mahavidyalaya, Kabi Surya Nagar, Orissa.**

of the country having a combined length 29,000 km, provide one of the richest fish genetic resources in world. The floodplains lake are primarily continuum of rivers Ganga and Berhamputra, these are as oxbow lakes (maunas, chaurs, Jheels, Bells has they are called locally, especially in Assam, Manipur, West Bengal, Bihar and Eastern U.P. reservoirs constituted the singles largest inland fishery resource both in source size and productive potential. According to FAO 1995 a total of 19370 reservoirs were found as small, medium and large resources in India with a total area of 3.15 million ha. Indian floodplain lakes (0.24 Million ha.) are primarily continuum of Ganga and Berhamputra. These are mainly located in Assam, W.B. Bihar and UP. They occupy important position inland fishery of India, because of their magnitude, their production potential serving as breeding and nursing ground for riverine fish stock. These water bodies are extremely rich in nutrients. The floodplain water land are lying uncared for and are in the process of swapmification. These are capable of yielding and average 1 tone ha, if subjected to scientific management. Therefore a vast untapped production potential is yet to be harnessed in the floodplain wetland.

Table 5.1: Water resource potential in India

Items	*Units*	*Quantity*
Length of the coastline	Km	8118
Exclusive economic zone	Million km²	2.02
Continental shelf area	000 km.	506
Length of rivers, canals K.M.		191,024
Brackishwater area	Million ha	1.23
Reservoirs area	Million ha	3.15
Tanks and ponds	Million ha	2.25
Beels, oxbow and derlict waters	Million ha	0.82
Marine fish potential	Millions tones	3.93
Indian fish potential	Million tones	
Marine		2.83
Inland		2.82
Total		**5.65**

Table 5.2: Resource potential of exclusive economic zone (000 tones)

Coast	*Demersal*	*Pelagic*	*Total*
North-west	756	461	1217
South-west	473	834	1307
North-east	142	178	320
South-east	313	241	554
Total			502
Lakshadweep oceans deep sea total			3900

With present day magnitude of 3.15 ha reservoirs from the important inland open water fishery resource of our country. The area under resources is expected reach 6.0 million ha in the another 20 years. The present fish production from reservoirs is estimated at 0.094 million tones with over 79 per cent contribution of small reservoirs follow by larges 14 per cent and medium (7%), at present level of management and utilization, they yield and average 20 kg./ha/year which is far below the potential, there for it characterised by low primary productivity of resources slow growth rate of fish in efficient fishing practice and in accessibility of fishing site.

Marine Fisheries Production

Marine capture fisheries play a vital role in India's economy, the sector provides employment and income to nearly 2 million people the growth in marine fish production during 1950s and 1970s was faster as compared to inland fisher is however, 1980 and 1990s, the trend was reversed, as the marine fish production was slow when compared with the inland fisheries. During the last decade 1990s the marine fish production reached a plateau. A quantity jump in catch from the presently exploited grounds and in certain areas up to 200 m depth is unlikely. Most of the major, commercially exploited stock are showing sign of over exploitation. But demand for seafood was growing. The current level of marine fish production from the exploited zone has to be sustained by closely monitoring the landing and the fishing, a effort and by strictly implementing the scientific management measure.

The share of inland fishery sector which was 29 per cent in 1951 has grown up to 49 per cent in 2001, indicating increasing contribution of inland sector to the total fish production. Fresh water aquaculture recourses of the country are estimated 30.67 million ha of which 2.25 million ha are ponds/tanks 8.27 million ha beels/jheels/derelict water bodies and 3.15 million ha reservoirs contribute 2.8 million tones production from these sources, cold water fishery resource encompass high and mid-altitude lakes, rivers, streams and man-made reserves. According to conservative estimates, the riverine stream length holding cold water fishery in the 2 mountainous zones trained by Indus and its tributaries, Jhelum, Chenab, Ravi, Veas, Satluj, Bhagirathi and Alkananda are about 7,000 kms. the approximate fishable stream length of hills stream on North-west and eastern Himalaya are estimated as 3200 km. further. There are large number of natural lakes located in the uplands of J&K Himachal Pradesh, Uttaranchal, Arunachal Pradesh and Sikkim, the natural eco-topes are virtual dotted all over the hilly terrain these states and hold rich ichthyofauna of both commercial and sport fishers. Coastal aquaculture production in the country largely on account of shrimp farming is estimated that out of about 1.3 million ha identifiably as potential areas for brackishwater fish farming in the country of the total area about 10 per cent area is being farmed at present of this area 80 per cent is under traditionally farming system. And the remaining is under extensive and semi-extensive shrimp farming. The activities of shrimp farming on the commercial scale are taken up in the Andhra Pradesh, Tamil Nadu, the traditionally farming system are located in W.B. and Kerala.

Table 5.3: Status and tend of fish production of India (Million metric tonnes)

Year	*Total Production*	*Marine Production*	*Inland Production*
1950-51	0.75	0.53	0.22
1961-62	1.16	0.88	0.28
1970-71	1.75	1.08	0.67
1980-81	2.44	1.55	0.89
1990-91	3.84	2.30	1.54
1999-2000	5.65	2.83	2.82

Table 5.4: Fish production in India

Year	*Marine*	*Inland*	*Total*
1980-81	15.55	8.87	24.42
1986-87	17.13	12.29	29.42
1990-91	23.00	15.36	38.36
1991-92	24.47	17.10	41.57
1992-93	25.76	17.89	43.65
1993-94	26.49	19.95	46.44
1994-95	26.92	20.97	47.89
1995-96	27.07	22.42	49.49
1996-97	29.67	23.81	53.48
1997-98	29.50	24.38	53.88
1998-99	26.96	25.66	52.62
1990-2000	31.22	27.19	58.41

Table 5.5: Fish production—projection for X five year plan

Base year	*Marine @2.5% growth rate/year*	*Inland @8% growth rate year*	*Total @5.44 growth rate/year*
2001-02	2.97	3.29	6.26
2002-03	3.04	3.555	6.59
2003-04	3.12	3.83	6.95
2004-05	3.20	4.14	7.34
2005-06	3.28	4.47	7.75
2006-07	3.36	4.83	8.19

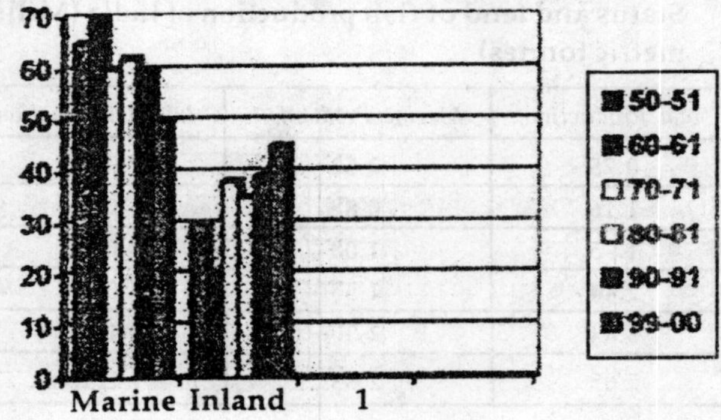

In the Contest of Silver Revolution

Fish is a good source of animal proteins. Man has realized its importance from the very inception of the evolution of the human race. It has been the sole diet for many island nations before the evolution of farming techniques. Even to day the fish supplies 35-60 per cent animals protein requirement of many Asian countries. It also plays a lead role in foreign trade. Around 180 countries trade marine products on a global basis. India's exports of marine products crossed Rs. 50000 million in 2001 and continue to grow at a rate of 10 per cent annually. From nutritional point of view, aqua food are unique as they provide not only superior quality proteins but also vitamins, calcium and phosphorus essential for health and well-being of people. In view of low energy yielding substance and the presence of essential amino acids, pol-unsaturated fatty acids rational, and be complex vitamins these food are plane vital role in the food and nutritional security system of the country for attaining the two objective of food and nutritional security for the present and future generation, we can examination in date the current and the future production scenarios and emerging challenges resulting form the unsustainable activities which have far reaching implication on aqua food production in the years to come. Under the changing scenario, we can opines that a pragmatic plan of action meets to be designed with definitive objective and activates by including protection and prevention components in the future aqua food production programme.

Table 5.6: Human population in the country

Year	*Total population*	*Average Annual Growth (%)*	*Person/Km²*
1951	361.0	1.26	117
1961	439.1	1.98	142
1971	548.2	2.20	178
1981	685.0	2.30	216
1991	846.3	2.50	267
2000	1400	–	–
2025	1400	–	–

Nutritional Status of Aqua Food

From nutritional point of view the aqua foods are valuable as they provide wide range of nutrients besides mineral in high proportion retinal, B group vitamins, calcium, and phosphorous for both poor and rich population of the country. The aqua food is a health food for the rural and urban masses of the country and therefore playing a crucial role to wards preventing, controlling and reusing the multiple nutritional deficiency diseases in the country.

Table 5.7: Nutrient content of common food fishes

Name of fish/ group	*Moisture*	*Protein*	*Fat*	*Carbohydrate*	*Calcium*	*Phosphorus*	*Iron*
Crops	78.0	20.0	1.4	3.1	462	300	0.9
Live fish	80.0	19.0	1.0	5.2	430	443	1.5
Catfish	70.0	19.2	6.4	2.3	270	170	2.0
Hilsa	54.0	22.0	19.4	2.9	180	280	2.1
Mahseer	78.0	18.0	1.9	2.0	110	80	3.8
Bhetki	80.0	15.0	0.8	3.0	480	350	3.1
Promphret	75.0	17.0	1.0	0.8	324	278	5.3
Mackeral	77.0	20.0	1.6	0.5	430	300	2.0
Mullet	70.0	19.0	7.8	1.5	357	175	4.4
Oil sardine	76.0	19.6	2.0	0.1	357	350	6.1
Shrimp	77.0	20.0	0.5		32	270	53

Roll of Fish Farming and Development Agencies in the Food Security

Fisheries sector occupy a very important place on the socio-economic development of the country. It has been recognised as a powerful income and employment generator as it stimulates growth of a number of subsidiary industries, and is a source of cheap nutritious food, besides being a foreign exchange earner. Most importantly, it is the livelihood of a large section of economically backward rural population of the country.

India is the sixth largest produce fish in the world and the second largest producer of inland fish the production of the country had increased from 2.959 million to 8.388 million from the year 1981 to 2001 at the annual average growth rate of 6.2 per cent, 2.95 yielding and 2.438 million tones. About 80 per cent of inland production reported from aqua culture. It estimated that the production potential of marine sector is 3.9 million and 2.9 million of inland sector, therefore fish farmers development agency has the mean objective of optimizing fish production productivity augmenting export of fish generating employment and improving welfare of the fisherman and their socio-economic status. In this regard:

1. the nature of fish seed production;
2. the nature of domestic and export fish marketing;
3. the nature of employment sector;
4. the nature of per-capita fish consumption have tremendously improved.

Strategy for Development in Fishery Sector

The main strategies and thrust area of future development of fishery sector are:

1. Expansion of aquaculture through upgradation of technology;
2. Conservation and optimum exploitation of the reverie fisheries;
3. Production of quality fish seed and feed for aquaculture;
4. Development of coldwater fisheries;
5. shrimp and prawn culture for meeting export requirements.

Fish as a Source of Employment

Fishing has been a major source of food for humanity from the tome immemorial. It not only provided food to man but also employment and economical benefits. Today around hundred million tones are caught by man from the seas to utilize them for alleviating hunger. Information of employment generation of fishery, the primary producer is scarce. However there are indications that their number is rapidly expanding in many developing countries like India, China, Thailand etc. in view of the high export trade and aquaculture. It is estimated that over 30 million youth are employed worldwide in the primary production of the fish and fishing operation. There are many ancillary industries supported by fishing operation. These include boat building, fishing gear manufacture, ice production, transport processing marketing and exporting other countries. On an average for every one youth employed in fishing operations 4 additional persons are getting support employment though these ancillary industries. But one major change that has taken place over these years is that fisheries generate employment for highly educated youths also. Modern fishing industry support employment of qualified skippers, boat-crews and also fishing technicians. Fish processing has given birth to a series of hi-tech hobs for youth. Modern seafood industries absorb processing technologies, microbiologists, quality control specialists as well as marketing managers. Fish farming in India offers an income generating highly remunerative for youth. It is a failure in philosophy if India planners think that high-tech industries are capable of generating employment for youth with the introduction of computer, robots and other fast production machine, modern industries hardly need much human labour these industry need high investment and each industry can at best hardly generate 1000 employment. But against the same investment in fishery sector there can be generation of for more employment opportunities with use of aqua food farming lands and population.

The Role of Fisheries in Rural Development

India ranks second in the world in the aquaculture production contributing about 7 per cent of total world production through aquaculture. Fishery sector in India has been providing employment to about 59.60 lakhs full time to part time fisher man in addition another six million people are engaged in fishery relation ancillary activates. And the sector contributes approximately 1.3 per cent of GDP through economic activates by small scale for much unemployed youth and women through income generation opportunities. In the contrast of rural development:

1. For sustaining rural livelihood, poverty eradication and food security aquaculture and its by product may be

undertaken as an economic activity by small scale for farmers, unemployed youth and women as source of income generation;

2. Awareness of utilization of available resources is the prime need amongst the fishery community;
3. The small-scale rural sector for the fishery and aquaculture requires more initial support and for poorer target groups, longer period are needed;
4. The extension and development approaches used for rural development with respect to fisherman need to be improved;
5. Through a holistic, farming system based on approached by integrating aquaculture in to rural livelihoods;
6. Participatory needed based approach that takes full account of the capacity of the poor the resources available to them and the risk they take to be taken into account;
7. The rural aquaculture has to be developed and promoted as an entrepreneurial activate that is financially viable even for small-scale operation should taken into account.

Conclusion

In the prospective of silver revolution the co-operative hands of both Govt. and private bodies. N.G.O. should come forward:

1. To stop environmental degradation at all cost;
2. Finance and financial requirement for fishery sector should enhance both at budgetary and non-budgetary provisions;
3. Modernization of fishery sector with the introduction hitch formula is required;
4. Cooperation and linkage between agriculture, industry and fishery sector is also required;
5. The prospective silver revolution should widen to rural sector for both income and employment generation, so that the reality of silver revolution will result as a revolution of mass benefit and a nationalistic in character.

6

Watershed and Rural Development

*Radhakrishna Panda**
*R.N. Dalai***

Abstract

Watershed approach for sustainable rural development has gained momentum in recent years. This programme is being implemented by involving the village people for their all round development. In Orissa, the programme is implemented in all the 30 districts under different schemes such as Integrated Wasteland Development Programme (IWDP), Drought Prone Areas Programme (DPAP), Employment Assurance Scheme (EAS), Revised Long Term Action Plan (RLTAP) (in KBK districts) etc.

The present paper attempts to analyse the socio-economic, environmental and other developmental changes in rural areas as a result of Janeinal watershed programme implemented in Keonjhar district under IWDP programme. The various quantitative changes are analysed on the basis of sample data obtained by employing quick appraisal techniques like Rapid Rural Appraisal (RRA) and focus group discussion. On the basis of the findings, generated in the study, the paper views that there has been sufficient socio-

* **Lecturer in Economics, P.G. Department of Economics, Khallikote (Autonomous) College, Berhampur, Orissa, 760 001. e-mail: ecorkp@yahoo.com**

** **Senior Lecturer in Economics, Khallikote (Auto) College, Berhampur, Orissa – 760 001.**

economic progress of rural people in terms of cropping pattern, cropping intensity, yield rate and gross revenue per hectare of land. Meat, milk and fish production tend to increase, human development indicators like increased enrolment ratio, reduced dropout ratio, lower incidence of LBW babies and increased institutional delivery etc. suggest successful changes associated with the watershed programme. Incidence of out-migration from the project areas is eliminated and food security promoted number of wells and tube wells owing to increased water table are reported to be increased.

Introduction

Rural development means sustained growth of income of the rural people along with the distribution of income and economic assets in favour of the poorer sections of the community in such as way as to progressively remove unemployment and poverty. With the progressive removal of poverty, rural development also brings about an improvement in social, educational, health, cultural aspects and other aspects of rural life (Mitra, 1985).

According to the World Bank (1975), Rural development is a strategy designed to improve the economic and social life of a specific group of people—the rural poor. It involves extending the benefits of development to the poorest among those who seek a livelihood in the rural areas. The group includes small farmers, marginal farmers, tenants and the landless labourers.

SECTION–I

Watershed Approach for Rural Development

Watershed have been playing a crucial role particularly for irrigation and multiple uses of mankind. The technique of watershed particularly the soil and moisture conservation techniques have become very much popular and successful to covert dry land (rain fed areas) into promising agricultural land. In the early twentieth century, attempts were made at ravine land reclamation in Gwalior state by constructing check dams and allowing the upstream area to silt up into level cultivated fields (Bali, 1990). Comprehensive watershed development programme carried out by Anna Hazare at Ralegan Sindhi (Ahmad Nagar district), Vilas Balwant Salunkhe's watershed programme at Niagaon (Pune district) and Vijoy Borde's

Adgaon watershed (Aurangabad district) in Maharashtra has converted poverty stricken, drought prone villages into lush green agricultural fields (Narayanan *et al.*, 1999).

In recent years efforts have been made in different parts of the country to develop improved measures to harvest the runoff water and recycle it to the cropped area when it is critically required. Thus, the harvested rain water could be utilised for double or multiple cropping in the areas which are otherwise conventionally mano cropping areas (Singh, 1990). Runoff water which, in earlier years were followed to flow in natural drainage system, now stored in tanks and reservoirs for irrigation. This has a great potential in establishing the dry land yields by way of collection of runoff water in dugout ponds and transferring the same back into the catchments areas during periods of moisture stress (Rao and Singh, 1990).

Since the inception of watershed management strategy in 1974, a number of watershed across the country have been materialised. Though the basic objective of watershed is soil and moisture conservation, simultaneously it influences the cropping pattern, yield rate of crops cultivated and consequently livelihood sustainability and food security of the people. In a study conducted on Munsiguda mini-watershed in Phulbani district of Orissa. It is found that due to this resource management programme, considerable progress has been achieved not only in crop production but also in improving the socio-economic condition of the inhabitants. The productivity levels of the traditionally grown crops like maize, ragi, black gram, green gram, mustard and groundnut have increased by 132.50, 38.75 and 70 per cent respectively (Bhol *et al.*, 1990). In another study on Padalsingi watershed project, it was observed that due to watershed, the per capita annual income of people in the catchments areas has increased from Rs. 1,587 to Rs. 6,541, the annual income per hectare increased from Rs. 1,780 to Rs. 7,815 and the cropping intensity increased from 106 to 150.7 per cent. The adoption of cropping pattern in the watershed catchments areas was to the extent of 90 per cent. The significant contribution of this project was inter-cropping of bajra and red gram which was to the extent of 100 hectares (Phadnmis *et al.*, 1990). The water harvesting structures (WHSs) created under IWDP programme at Kashipur have positively influenced cropping pattern by means of crop shifting from traditional crops to more valuable cash crops. Perceptible changes are noticed in the yield rate of almost all the crops and the gross return per hectare of land is found satisfactory (Panda *et al.* 2005).

Janeinal Micro Watershed Project

During the period 1998-99 to 2005-06, with the objective of increasing agricultural production and simultaneously improving rural development, Janeinal Watershed Project had been implemented under IWDP-I in Keonjhar district. The project is being implemented by the Soil Conservation Department, Keonjhar. IWDP-I, consists of 11 micro-watersheds covering 27 villages of which 25 villages are under Patna Block and the rest 2 are at Keonjhar Block.

The project area comes under North Central Plateau and North Eastern Coastal plains agro-climatic zones. Red soil and coastal alluvial soils constitute the soil structure. The mean temperature varies in the range of 8° in winter and 38° in summer. Topographically, this area is undulating with crop lands and developing in terraces. The process of denudation is accelerating resulting in gradual depletion in the forest area. In this background, with the objective of preventing large-scale denudation, protecting soil loss, enhancing moisture stress, improving agricultural productivity and accessing to food, fuel, fodder and fibre for the community the IWDP (1) has been introduced in this area.

Objectives of the Study

The present paper attempts at conceptualising Rural Development and Watershed approach for rural development in Section-1. In Section-2, the implementation procedure of Janeinal watershed is highlighted by analysing different components of Janeinal Watershed programme. Considering a number of socio-economic environmental and rural development parameters, the impact of watershed programme on sustainable rural development is analysed in Section-3.

Study Approach and Methodology

The impact analysis is done by comparing the selected indicators during the post-project period with the corresponding values during the pre-project period i.e. 1 year preceding the year in which I.W.D.P. intervention was started. For assessing the type of change, only Balaniposi micro-watershed out of 11 micro-watersheds were selected randomly and data were obtained by means of Quick Appraisal Techniques. Section-I, Section-II of this study is mainly based upon secondary data collected from published books, journals and the official reports of A.S.C.O, Keonjhar. On the other hand, the analysis in the Section 3 is based upon the results obtained from Quick Appraisal Techniques basing upon Focus Group Discussion and Rapid Rural Appraisal.

SECTION-II

Project Components

Project components are analysed in terms of treated area as a proportion to treatable area, financial target and achievements, physical target and achievements. It is reported that in the project area, all the treatable area constituting 5145 hectares have been treated. Of the geographical area of 8237.29 hectares, around 62 per cent are treated in the project area under IWDP (1). Out of the total funds of Rs. 205.17 lac received by the PIA, Rs. 203.52 lacs are utilised during the project period 1998-99 to 2005-06. The items of work consist of entry point activities and assets created out of WC funds. Entry point funds is mainly temple boundary walls, and community choupals used for conducting watershed meetings. Gully-plugging, field bonding, SR Dam, farm pond, check dam, percolation tanks, run-off management structures etc. are created out of the WC funds. Besides, the activities like renovation of existing structures, are also taken up. The project constitutes with 571 user groups consisting of 4024 persons. The members of the user groups are contributing 10 per cent of the respective project cost of WDF account. However, SC and ST beneficiaries are contributing only 5 per cent of the project cost. There are 109 self help groups consisting of 1292 members have been facilitated under the watershed programme. The total saving mobilised by all the SHGs stand at Rs. 5.46 lakh and 87 per cent of these SHGs are linked with bank, it is found that the SHG members are mostly engaged in rice business and vegetable production/vending. In most cases SHG members are marginal farmers and landless agriculture and labourers.

It is reported that out of 11 micro watershed, plantation programme is undertaken in 10 micro watershed. In all these micro watershed plantation activities are done in 175.45 hectares of land. At the early stage of plantation, nursery was prepared for the watershed reporting plantation. Cashew Simorouba, Acacia are the different plantation species reported. The rate of causality of plants varies in the range of 10 per cent -35 per cent. The different personnel associated with the implementation and watershed programme and the members of the user groups have received different training programme. It all the micro watersheds, two bank accounts like WDF A/C and WC A/C are maintained. The total savings in WDFS A/c stands at Rs. 821638/.

SECTION-III

In this section the impact of watershed programmes of socio-economic, rural environmental developmental is analysed on the basis of quick appraisal techniques involving focus group discussions and rapid rural appraisal.

Socio-economic Impact

The socio-economic impact of watershed is analysed by comparing the present status of different indicators during 2005-06 (post project period) with the corresponding figures in the year 1997-98 (pre-project period). The marginal charges are traced not by calculating the percentage deviation.

The different socio-economic indicators include cropping pattern, yield rate of only paddy, cropping intensity, gross revenue, livestock population, fishery, enrolment ratio, dropout rate, incidence of infant mortality rate, incidence of low birth weight babies, incidence of out migration, incidence of food insecurity (hunger), institutional delivery etc. The socio-economic impact of watershed programme is shown in Table 6.1. It is revealed that due to WHSs, there has been crop shifting from traditional cereals and vegetables to high value off-season vegetables. Due to crop demonstration programmes, farmers have introduced new vegetables. Yield rate of paddy has increased by 71.42 per cent. Cropping intensity has increased from 66.66 per cent during pre-project period to 134.33 per cent. There has been spectacular increase in the Gross Revenue per hectare. Consequent upon increased agricultural production, and availability of fodder, livestock population has increased. The percentage increase in the number of cows, buffaloes, goats and sheep are 7.13, 33.35, 50.00 and 66.66 per cent respectively. Milk production in the study area tends to increase because of the increased number of cows. Similarly, more number of goats and sheep indicate increased meat production.

Case Study 1

Watershed Programme has influenced cropping pattern

Mr. Kailash Chandra Mohant, aged about 50 years, an inhabitant Balaniposi micro watershed is a farmer, owning 9 acres of land. Before IWDP intervened, around 40 per cent of his land holding were wastelands. Now, these wastelands are converted into cultivable lands due to the irrigation potentials created by the WHSs. Now, he is able to produce a large variety of off-season vegetables and sells at a higher market price. Under plantation programme, he has introduced mango orchards, cashewnut orchards. From the plantation of Seimarao trees, he is able to produce oilseeds. Before IWDP project, paddy was the major crops produced by the farmers. But, now in addition to the increased paddy production, he is able to produce vegetables, fruits and oilseeds. He says that his annual income has increased from Rs. 10,000 to Rs. 1,00,000 per annum from his agricultural activities. With the increased income, he has renovated his own house, and has saved Rs. 30,000 at the nearby bank which was a dream for him during the pre-project period.

Case Study 2

Watershed programmes symbolises rural development.

Madan Mohanty, aged about 35 years, an enterprising farmer owns only 1.25 acres of land. Soon after the creation of WHSs close to his farm, he was encouraged enough to produce off-season vegetable for the sake of higher profit. After his first experiment, he cultivated more knowledge to get good profit by selling more off-season vegetables. In the subsequent years, he undertook such firming on lease basis. Now, with his 5 acre land holdings including his leased in lands, he produces varieties of vegetables like cabbage, papaya, brinjal, green leafs, tomato, radish, bitter-ground, pumpkin, cucumber, bottle-gourd etc. Now the farmer says, "Due to irrigation potential of WHS, created under IWDP programme, he has been able to get Rs. 1-1.5 lakh income per annum."

Madan Mohanty at his own pucca house

With his increased income he has constructed a pucca house, having a solar lighting system, owns a motor cycle and watches television run by a battery system. This is the real success of watershed programme and Madan Mohanty symbolises the rural development due to watershed programme.

Similarly, fish production in the project area has increased by 50 per cent. In this background, in the project area, sufficient crop-diversification is materialised. So far as human development aspects are concerned, enrolment ratio is increased, dropout rate is reduced to the extent of 85.00 incidence of LBW babies are not reported due to the Health Awareness camps. Out-migration of labour is totally prevented because of increased and remunerative on-farm employment opportunities available at own place. Incidence of hunger and food-insecurity is lowered. Institutional delivery of pregnant women is increased from 5 per cent to 20 per cent. Thus, watershed programmes have tremendously influenced the socio-economic life of rural people.

The environmental impact of watershed programmes are assessed by considering the indicators as mentioned in the Table 6.2. It is found that no. of wells and tube wells are increased by 116.66 per cent and 500.00 per cent respectively. There has been improvement in water table. However maximum temperature during summer tends to increase. It is because of the increased industrial and mining activities in the nearby areas.

Table 6.1: Socio-economic impacts of watershed programme

Sl. No.	*Indicators*	*Pre-project*	*Post-project*	*% of deviation*
1.	Cropping pattern (No. of crops)	5 (paddy, brinjal, gourds, lady finger, cabbage	(Paddy, brinjal, gourds, lady finger, cabbage potato, radish, horsegram, maize, tomato, til, sunflower peanut and beans	120.00
2.	Yield rate of paddy in qtl/Acre	7	12	71.42
3.	Cropping intensity (GCA as % to land holding)	66.66	134.33	101.51
4.	Gross revenue per acre	Rs. 8,000	30,000	275
5.	Livestock population No. in the sample village			
	Cows	140	150	7.14
	Buffaloes	30	40	33.35
	Goats	40	60	50.00
	Sheep	3	50	66.66
6.	Mil production (litre of milk produced per month)	150	200	33.33
7.	Fish production	100	200	100.00

(Table 6.1 Contd...)

Sl. No.	*Indicators*	*Pre-project*	*Post-project*	*% of deviation*
8.	Enrolment ratio (5-14 years children admitted in schools)	60	90	50.00
9.	Dropout rate (%) of children of 5-14 years, joined in school, but left between the session	20	3	-85.00
10.	Incidence of LBW babies (Nos. less than 2500 gms.) during birth	2	0	–
11.	Incidence of institutional delivery (%) of pregnant women undertake delivery at hospitals (clinics)	5	20	300.00
12.	Incidence of out migration (No. of persons reporting in a year)	20	0	–
13.	Incidence of hunger (No. of HHs remaining hungry for few days during a year)	25	2	-92

Impact of Watershed Programmes on Rural Development

In the sample village, there were no pucca house during the pre-project period. Now-a-days, there are about 10 household who are having pucca houses due to increased agricultural income and agro-diversification. There has not been any change in the number of wells inside the house, though wells for agricultural purposes have increased as per Table 6.2 Access to latrine facility is not reported in any of the households. No. of motorcycles has increased by 350 per cent. No. of cycles in the project village has increased. Use of gas stoves are reported among few households which was nil during pre-project period. Thus, even though there is poor sanitation, in other way rural development is definitely taking place, on the basis of the socio-economic changes as per Table 6.1 and 6.2.

Table 6.2: Environmental impacts of watershed programme

Sl. No.	*Indicators*	*Pre-project*	*Post-project*	*% of Deviation*
1.	No. of wells in two sample villages	30	65	116.66
2.	No. of tube wells	2	12	500.00
3.	Water table	Below 24 ft.	Below 12 ft.	-50.00
4.	Maximum temperature during summer	38	4	18.42

Concluding Remarks

Due to watershed programmes, there has been sufficient socio-economic progress of rural people in terms of increased income, which is due to the changed cropping pattern, better cropping intensity, yield rate and gross revenue per hectare of land. Increase in meat, fish, milk production is also reported. Human development indicators like increased enrolment ratio, reduced dropout ratio, lower incidence LBW babies, and increased institutional delivery suggest successful changes associated with the watershed programme. The incidence of out-migration is eliminated and food security is promoted. The quality of environment is increased due to moisture conservation. As a result of which more number of wells, and tube wells are installed in the project area. There has been improvement in the water table. However, maximum temperature during summer tends to increase because of the increased industrialisation and mining operations in the nearby project areas. Except sanitation there has been many sided rural development in the project area.

Table 6.3: Impact of watershed on rural development

Sl No.	*Indicators*	*Pre-project*	*Post-project*	*% of Deviation*
1.	No. of pucca houses (in one sample village)	0	10	–
2.	Wells inside house (in 1 sample village)	16	16	0
3.	Access to latrine facility	0	0	0
4.	No. of motor cycles	2	9	350.00
5.	No. of cycles	60	100	66.66
6.	No. of households possessing gas stoves	0	4	–

Post-project interventions like hand over of difficult assets to the community should be ensured. A fresh training programme, might be arranged for the watershed committee members regarding the sustainability of the project. Few assets like plantation in community lands and some water harvesting structures should be handed over to the SHGs, considering the views of the community. There should be convergence of district WATSN programme with the watershed villages during post-project period, this might lead to increased access to sanitation. Because of tremendous socio-economic, environmental and rural development impacts of watershed programmes the intervention procedure of the concerned PIA is excellent.

REFERENCES

1. Bali, J.S. (1990) *Integrated Watershed Management: A National Perspective,* International Symposium on Water Erosion, Sedimentation and Resource Conservation, October 9-13, Dehradun, India.

2. Bhol, B.B., Senapti, P.C. and Dikshit U.N. (1990), *Resource Management in Rain Fed Watershed: A Case Study of Munsiguda Mini Watershed in Phulbani District of Orissa,* International Symposium on Water Erosion, Sedimentation and Resource Conservation, October 9-13, Dehradun, India.

3. Dhabriya S.S. (1999), *Desert, Drought and Development: Studies on Resource Management and Sustainability*, Hooja and Joshi (Ed./S), Rawat Publications, New Delhi, p. 298.

4. Foster, J., Greer. J. and Thorbecke, E. (1984): *A Case of Decomposable Poverty Measures*, Econometrica, Vol. 52 (3), pp. 761-766.

5. Jain, H.K. (1985): *Indian Agriculture in 2000 A.D.*, Indian National Science Academy, New Delhi.

6. M.S. Swaminathan Research Foundation (2001): *Food Insecurity Atlas in India*.

7. Mitra, Ajit Kumar (1998): Theories of Rural Development Implicit in the Rural Development Programmes Implemented in India in the Post-Independence Period, in Sachinandan Sau (ed.), *Rural Development Theories and Experiments*, Allied Publishers Limited, New Delhi.

8. National Bureau of Soil Survey and Land Use Planning (2001): *Classification on Dry Land Regions in India*, Nagpur.

9. Panda, B.K., Panda R.K. and Sarangi P. (2005): "Impact of Water Harvesting Structures on Cropping Pattern and Food Security, *Orissa Economic Journal*, Vol. 37 (1 & 2).

10. Paul, J.C. and Sharma, S.D. (1999). Micro Irrigation for Water Scarce Areas of Orissa, *Kurukshetra*, Vol. 47 (9).

11. Phadnawis, A.N. Birajdar, N.M., Nanagre, M.L. and Asmatoddin Md. (1990), *Integrated Management of Resources of Watershed Basis for Optimising Production—Padalsing Watershed*. International Symposium of Water Erosion, Sedimentation and Resource Conservation, October 9-13, Dehradun, India.

12. Ramakrishna, Y.S. and Venketshwarlu, (2004): Global Opportunities Galore, Survey of Indian Agriculture, *The Hindu*.

13. Rao, M.S. Ramamohan, Chittaranjan S. and Singh Gurmel., I (1990): *Soil and Water Conservation Measures for Red and Black Soils of Semi Arid Region of South India*, International Symposium on Water Erosion, Sedimentation and Resource Conservation, October 9-13, Dehradun, India.

14. Shah, Amita and Guru, Baidyanath., (2003), *Poverty in Remote Rural Areas in India: A Review of Evidence and Issues*, Gujarat Institute of Development Research, Working Paper No. 139.

15. Sharma, Devinder, (2003), *Bio-tech Age: From Hunger to Hidden Hunger*, Financing Agriculture: Vol. 35 (4).

16. Shiva, Vandana, (2003), *Having Our Cake and Eating It Too (Conserving Bio-diversity While Feeding the World)* Financing Agriculture, Vol. 33 (4).

17. Singh Gurmel, (1990), *Rain Water Harvesting and Recycling for Sustainable Agricultural Production*, International Symposium on Water Erosion, Sedimentation and Resource Conservation, October 9-13, Dehradun, India.

18. Singh, H.P. Sharma, K.K., Korwar, G.R. (2003), *Sustainable Dry Land Agriculture Development*, Financing Agriculture, Vol. 35(1).

19. Sivanappan, R.K. (2004), *Water Harvesting and Conservation in Dry Land on Watershed Basis*, Financing Agriculture, Vol. 36(4).

20. World Bank (1975), *Rural Development Sector* Policy Paper, Washington DC.

7

Socio-economic and Environmental Changes and Joint Forest Management

*Radhakrishna Panda**
*Jagannath Panda***

Abstract

The governance of forest resources through Joint Forest Management is of emerging significance. Orissa is among the 27 states in India to implement this programme since last several years. The main provisions of Joint Forest Management include maintenance of environmental stability, conservation of soil/ moisture, increase in forest productivity and sustainable increase of forest resources. Present paper attempts to assess the socio-economic and environmental changes due to the changed governance like Joint Forest Management at Gundibira village under South Ghumsar Division in Orissa. In this village a JFM was constituted in the year 2002-03 to protect 185 hectares of AR plantation and 15 hectares of ANR plantation.

Data on different indicators pertaining to socio-economic and environmental effects of JFM were obtained with quick appraisal methods. The techniques like focus group discussion, observation method and rapid rural appraisal methods were employed. Analysis

* Lecturer in Economics, Khallikote (Auto) College, Berhampur, Ganjam, Orissa, (ecorkp@yahoo.com)

** President, Joint Forest Management Committee, Gundibira, Sorada, Ganjam, Orissa, 761 108.

of data reveals that the introduction of JFM has positively influenced the socio-economic life of the rural people. Landless agricultural labourers have marginally escaped from poverty. Milk production and meat production in the VSS areas have increased. The conservation of soil/moisture is satisfactorily improved. Groundwater table is increased. Forest firing is prevented and bio-diversity enhanced. People's involvement in the preservation and protection of forests is very much proactive. The paper views that there is a tremendous scope for socio-economic betterment and environmental prosperity through Joint Forest Management (JFM).

Introduction

In recent years, there has been an interesting intervention in the governance of forest resources through Joint Forest Management. Significant progress of this type of governance is reported in terms of area under JFM and the number of associated Forest Protection Committees. As per the recent information available, 27 State Governments have issued resolutions in favour of JFM for the protection and conservation of forest resources. By December, 2002 there were 64,000 JFM committees managing 14.26 million hectares of forest land in different states (MOEF, 2003).

The development of Joint Forest Management owes much to the National Forest Policy, 1988, which emphasises the involvement of village committees living close to the forest in protection and development of forests. The subsequent notification in 1990 by the Government of India to State Governments makes an instruction to involve local communities in the management, protection and development of forests. In the year 1995, the Resolution was slightly modified to involve women and landless households as well as usufruct sharing by the communities. To provide legal backup to JFM committees, it was further revised on February 21, 2000. The main provisions of JFMs, as per, forest policy 1988 includes:

(i) Maintenance of environmental stability through preservation and restoration of ecological balance;

(ii) Conservation of natural heritage;

(iii) Check on soil erosion;

(iv) Sustainable increase in forest coverage through massive afforestation;

(v) increase in forest productivity and encouragement for massive utilisation of forest resources.

Functioning of JFMs

The Joint Forest Management needs a Village Level Organisation (VLO) of people for the purpose of their participation in the management. Such organisations may be the existing village Panchayat itself, or may be a newly formed organisation such as a cooperative society or a development society or a forest protection committee. Actually, most of the VLOs involved in JFM are in the form of Village Forest Protection Committees. The primary objective of village Forest Protection Committee is to provide a visible role to the local communities in planning, management and protection of forests and to give them a share in the benefits from these forests. Thus, the basic thrust of JFM is to establish grassroots community based institutions for the protection and management of forests.

JFM in Orissa

Presently, Orissa is among the 27 states to implement the JFM programme. As on 1st January 2000, in Orissa, 3704 JFM committees had been formed accounting to 4,19,306 hectares of forest land. Orissa is the third highest state implementing JFM programme more vigorously after Madhya Pradesh, and Andhra Pradesh (Planning Commission, 2001). The village forest Protection Committee in Orissa is called 'Vana Surakhya Samitee'.

JFM at Gundibira Village

At Gundibira village, under Ghumsar South Division, a Forest Protection Committee was found in the year 2002-03. The main purpose of this VSS is to conserve 185 hectares of Artificial Regeneration (AR) plantation consisting of 1,20,000 teak trees and creation and maintenance of 15 hectares of Added Natural Regeneration (ANR) plantation consisting of 4,500 teak trees. Of the targeted financial outlay of Rs. 1,43,813 by January 2006, Rs. 1,30,200 has been released in favour of the VSS under study (the details of financial target and achievement is shown in Annexure-II). The released amount has been utilised for the protection of 185 hectares of AR plantation and 15 hectares of ANR plantation by involving 1140 mandays labour for forest cleaning and maintenance. Out of the sanctioned amount, one water harvesting structure (WHS) in the forest has been created. Besides, to win community support 2 kms. *kutcha* road in the village, maintenance of the existing community hall and community utensils have been provided (the details of the physical achievement in Annexure-II). It is reported that at present all the villagers are committed towards the protection and conservation of the village forests.

Objectives and Methodology

The present paper is a modest attempt to analyse the socio-economic and environmental changes occurred due to the JFM activities in afforestation, conservation and protection of forests at Gundibira Village under Ghumsar (South) Forest Division in Berhampur Circle.

The findings of the present study are outlined based on the primary data obtained through Focus Group Discussions and Rapid Rural Appraisal (RRA) Methods. Since the year 2000, preservation and protection of forests by the village Forest Development Committee (Vana Surakhya Samitee) is the major environmental intervention programme carried out in this village. The various socio-economic and environmental changes taking place during this period has been viewed as the outcome of JFM. The changes in the socio-economic and environmental parameters are measured by comparing the present situation (Post-JFM) with that of corresponding figures before JFM (Pre-JFM).

Socio-economic Impacts of JFM

To examine the socio-economic changes due to Joint Forest Management, the indicators like incidence of small farmers, marginal farmers, agricultural labourers, forest dependent households and the type of migration etc. are analysed. The details of the socio-economic indicators and their changes due to the present JFM is shown in Table 7.1.

Table 7.1: Socio-economic changes due to JFM

Sl. No.	*Indicators*	*Pre-VSS*	*Post-VSS*	*Net change (% deviation)*
1.	Incidence of small farmers (per cent of households to the total (households)	19.08	19.08	Nil
2.	Incidence of marginal farmers (per cent of households to the total households)	28.24	42.74	51.25
3.	Incidence of agricultural labour households (per cent of households to the total households)	48.85	34.35	-29.66
4.	Incidence of forest dependent households (per cent of households to the total households)	38.16	3.81	-90.01
5.	Migration of adult labourers (No. of persons)	25	55	120.00
6.	Migration of child labourers (No. of child labourers)	15	6	-60.00
7.	Dropout rate	40.81	12.72	-68.83

(Table 7.1 Contd...)

Sl. No.	*Indicators*	*Pre-VSS*	*Post-VSS*	*Net change (% deviation)*
8.	Low birth weight babies/ Year (average separately calculated during Pre-VSS and Post-VSS	1.5	1.0	-33.33
9.	Livestock population (in Nos.)			
	No. of cows	300	500	66.66
	No. of bullocks	12	60	400.00
	No. of goats/sheep	60	200	233.33
10.	Average milk production (litres/month)	900	1500	66.66
11.	Shops (Nos.)	2	5	150.00
12.	Artisan households (Nos.)	5	10	50.00
13.	Women entrepreneurship (Nos. of SHGs)	0	6	–

Source: Computed from the field data.

It is revealed from Table 7.1 that during the post-VSS period, the incidence of small farmers in the VSS areas remaining unchanged, the incidence of marginal farmers has increased by 51.25 per cent and the incidence of agricultural labour households is reduced to the extent of around 30 per cent. This implies that most of the agricultural labour households during the pre-VSS period have moved upward and become marginal farmers. Similarly, the incidence of forest dependent households, those who are exclusively dependent on forests for their livelihood, is significantly reduced and the extent of reduction is around 90 per cent. It is inferred that there has been better occupational mobility. Indeed, among these households, sufficient awareness for the protection of forests are observed. Besides, these households are also pressurised to leave illicit forest activities, particularly timber collection and sales for some better economic activity. As a result, out-migration from the VSS area to Surat textile mills has increased to the extent of 120 per cent. Similarly, livestock activities during the post-VSS period have also increased sharply. The increase in the livestock population such as the number of cows, bullocks, goats and sheep has increased by 66.66 per cent, 400.00 per cent and 233.33 per cent respectively. During the Pre-VSS period, there were no *Holstein* cows in the VSS area, but during the Post-VSS period, around 10 *Holstein* cows are reported.

The average milk production during a month in the VSS area has increased by 66.66 per cent. The increase in the number of shops, artisan households, and women entrepreneurship is also found favourable. Previously, there were no SHG in the VSS-area, but now-a-days there are 6 enterprising women SHGs operating in the village. Seasons greetings, and calendars manufactured by these SHGs in collaboration with the respective JFM are even exported to foreign countries.

Even the human development indicators like dropout rate[1] and incidence of low birth weight babies[2] has reduced. This is due to the increased awareness consequent upon the health camps and literacy campaigns periodically conducted by the Forest department people and the concerned VSS.

Environmental Impacts of JFM

The environmental impacts arising due to JFM are examined by analysing different indicators associated with soil conservation, moisture conservation and improved environment. Indicators like no. of tube wells, no. of dug-wells, both type of wells remaining dry during summer, mean temperature during summer etc. have been studied to understand the moisture stress created by the adjoining forestry presently managed by VSS. It is revealed from Table 7.2 that though these has not been any increase in the number of tube-wells, but the number of dug-wells is increased by 14.28 per cent. The number of tube-wells is not increased because tube-wells are installed by the government departments. Similarly, the number of tubewells and well remaining dry during summer are reported to have been reduced to the extent of 50 per cent and 60 per cent respectively. Improvement in the water table to the extent of 25 per cent is also reported. Thus, overall performance with respect to moisture conservation is found satisfactory. Mean temperature during peak summer is also reported to be declining from 45° to 40°.

For assessing soil conservation, indicators like amount of sand and stone transported from VSS-site to construction sites are studied. Due to the active involvement of VSS people, such type of transportation has been checked 100 per cent. As a result, soil erosion is prevented. This is a good outcome of VSS.

Environmental protection due to VSS efforts are analysed by studying the variables like amount of timber transported from the forest areas to outside, number of days of forest firing during summer, number of picnics arranged at VSS forest. It is found that during the post-VSS period, there has been around 99 per cent reduction in the amount of timber transported and 90 per cent reduction in forest firing. Almost 650 per cent increase in the number of picnics arranged at VSS sites is also reported. Thus, in this background the environmental promotional role of VSS is profound. Effective utilisation of minor forest produces (MFPs) is also one of the important objectives of JFM. It is reported that in the study area, MFPs are not found much. However, it is found that the number of households engaged in leaf-plate making is sufficiently increased and the increase is as high as 400 per cent.

Table 7.2: Socio-economic changes due to JFM

Sl. No.	Indicators	Pre-VSS	Post-VSS	Net change (% deviation)
1.	No. of tube wells	6	6	0
2.	No. of wells	7	8	14.28
3.	No. of tube wells remaining dry during summer	2	1	-50.00
4.	No. of well remaining dry during summer	8	5	-60.00
5.	Water table during summer	25′	20′	-25.00
6.	No. of tourists to visit VS site	0	10	–
7.	Mean temperature	43°	40°	12.5
8.	Stone transported (No. of truckloads/week)	20	0	-100.00
9.	Red sand transported (No. of truckloads/week)	8	0	-100.00
10.	Timber transported (No. of bicycle-loads/week)	500	2	-99.6
11.	Households engaged in leaf plate making	2	10	400.00
12.	Picnics arranged at VSS site	2	15	650.00
13.	Forest firing (No. of days during summer)	20	2	-90.00

Source: Computed from the field data.

Conducting Remarks

With the introduction of JFM for preservation, protection and maintenance of forest resources, there has been a marked improvement in the socio-economic fabric of the rural people. There has been marginal escape from poverty among the landless agricultural labourers. Milk and meat production in the rural areas have increased. There has been satisfactory soil and moisture conservation due to the forest protection and afforestation activities undertaken by the VSS. Added to such benefits, ground water table has improved, forest firing prevented, bio-diversity in VSS areas enhanced and occupational mobility improved to a satisfactory level. Effective utilisation of MFPs is also reported. People's involvement in the protection of forests is tremendously proactive. However, issues like delays in sanction and insufficient provisions of training and exposure visits for the members of the JFM has partially hindered the progress. Considering that there is tremendous scope for the governance of forest resources through JFM all types of constraints associated with such type of governance should be removed.

ANNEXURE-I

Forest Circles in Orissa	Angul, Berhampur, Koraput, Sambalpur, PDSTR
Forest Divisions in Orissa	Angul, Athagarh, Athmallik, Baliguda, Bonai, Boudh, Bolangir, Deogarh, Dhenkanal, Ghumsar (North), Ghumsar (South), Jeypore, Kalahandi, Khariar, Nayagarh, Paralakhemundi, Phulbani, Puri, Rairakhol, Rayagada, Sambalpur, Sundergarh.
Forest Divisions under Berhampur Circle	Baliguda, Boudh, Ghumsar (North), Ghumsar (South), Nayagarh, Paralakhemundi, Phulbani, Puri.

ANNEXURE-II

Financial Targets and Achievements

Sl. No.	Years	Financial targets for different items of work			Total
		Advance work	Creation work	Maintenance work	
1.	2002-03	21,930	–	–	21 930
2.	2003-04	–	31,695	–	31,695
3.	2004-05	–	–	16,575	16,575
4.	2005-06	–	–	14,625	14,625
5.	2006-07	–	–	14,675	14,675
	Total	**21,930**	**31,695**	**43,875**	**97,500**

Financial Achievements

Sl. No.	Years	Funds received by the VSS			Total
		Advance/creation of work	Maintenance	EP	
1.	2002-03	–	–	–	–
2.	2003-04	53,625	16,575	–	60,000
3.	2004-05	–	–	–	–
4.	2005-06	–	–	–	–

ANNEXURE-III

Physical Target and Achievements

Sl. No.	Items	Target	Achievements
1.	Treatable area (hectares)	15	15
2.	Man-days employment	2865	1140
3.	WHS (No.)	1	1
4.	Village road (Km. of Kutha road)	2	2
5.	Maintenance of community hall (in Rs.)	20,000	20,000
6.	Provision of community utensils (in Rs.)	12,000	12,000

Notes

1. Out of the 1 enrolled children in primary schools, proportion of children have left school in between the education sessions. The mean dropout rate of the preceding five years to the introduction of VSS and succeeding 3 years after the India of VSS have been calculated.
2. Babies weighing less than 2,500 gms. during birth are considered are considered LBW (Low Birth Weights).

REFERENCES

1. B. Pramod and N. Murari Ballal (1992), *Sustainable Development of the Rural Poor*, Pigmy Economic Review, Vol. 38, No. 1.
2. Guumsar South (2004), *Unpublished Official Records*, Government of Orissa.
3. Ghumsar South (2005): *Pratibimba, Half-yearly Newsletter of FDA*, Ghumsar South Forest Division, Issue I.
4. Saxena Rakesh (2004), Governance of Forest Resources in India: A Brief Review and Identification of Current Issues, *Paper Presented at Eastern Zone Reflective Workshop on Governance: Gender, Poverty and Livelihood* held at Nabakrushna Choudhury Centre for Development Studies, Bhubaneswar, 20-21, July.
5. Panigrahy, Srikant (2004), *Sustainable Resource Use Through Participation*, Yojana, Vol. 48.
6. Planning Commission (2002): *Orissa Development Report*, Government of India.
7. Bellamy Carrol (2005): *The State of the World's Children*, United Nations Children's Fund.

8

Infrastructure

The Thrust Area of Rural Development

Dr. Adity Kumar Patra*

Infrastructure is a crucial input for overall economic development. Without having a sound infrastructure base a country cannot develop its economy. Infrastructure sectors are the backbone of a national economy. The word 'infrastructure' is defined in dictionary as "the underlying foundation or basic framework". By this definition, infrastructure is the basis for development. Infrastructure is considered as the foundation on which the factors of production interact in order to produce output. In the general sense, infrastructure refers to those basic facilities and services, which facilitate different economic activities and thereby help in economic development of the country.

Infrastructure can be classified as 'Economic Infrastructure' and 'Social Infrastructure'. The former includes transport and communication, energy, banking and insurance etc., whereas the latter includes sectors like health, education, housing etc.

There are five important characteristics of infrastructure services. They are:

1. Infrastructure services are monopolistic in nature;
2. They usually involve high upfront costs and long payback periods;

* **Lecturer in Economics, Kalinga Mahavidyalaya, G. Udyagiri, Kandhamal, Orissa.**

3. Investments are typically bulky and lumpy;
4. These services are marked by presence of externalities. This reveal the 'Public Good' nature of infrastructure, i.e. the characteristics of 'non-rival' and 'non-exclusive';
5. These involve high volume of 'sunk costs'.

Thus, high volume of sunk costs, long periods of gestation, inability to levy of user charges and prolonged project development makes the infrastructure sector investment highly risky. Consequently, infrastructure services have been mainly provided by the public sector in all most all countries during 20th century.

The availability of infrastructure facilities helps the economy to increase production, expanding trade, reducing poverty, improving the quality of life and better environmental conditions.

Infrastructure contributes to development both directly and indirectly. The linkage between economic infrastructure and development are as follows:

1. Infrastructure lowers the cost of producing a given level of output of alternatively, can increase the amount of output produced by all other inputs for a given cost;
2. Infrastructure enables markets to work better. Transactions are made less costly and this increases the benefits of trade;
3. Unit costs tend to rise due to unreliable and/or inaccessible public infrastructure. Firms spend a significant portion of their expenditure on buying infrastructure services and suffer when these are not available.

India is predominantly a rural economy. Three-fourths of people of India live in rural areas. The rural India is characterized by high incidence of poverty, widespread unemployment and under-employment and lack of infrastructural facilities. Rural development is the number one challenge before the nation. It has been the main thrust of planners and policy makers ever since independence. During the last five and half decades of planning era a number of policies and programmes were launched to strengthen the rural economy by creating a good infrastructure base.

The analysis of incidence of poverty across Indian states indicate that poverty is very costly linked to the absence of economic and social infrastructure. Investment in rural infrastructure have been reported to have a significant impact on increase in economic activities in the area and consequent reduction in poverty. Availability of infrastructure has also gone a long way in redressing the issues of regional imbalance.

The Eleventh Finance Commission had prepared an 'Index of infrastructure' by taking into account economic, social and administrative infrastructure. If we put the index of infrastructure vis-à-vis the incidence of poverty we find that states with high value of infrastructure index have low incidence of poverty. This is reflected from Table 8.1.

Table 8.1: Index of infrastructure and incidence of poverty (1999-2000 in %)

State	*Index of infrastructure*	*Incidence of poverty*
Andhra Pradesh	103.30(9)	15.77(6)
Assam	77.72(13)	36.09(12)
Bihar	81.33(11)	42.60(14)
Gujarat	124.31(5)	14.07(4)
Haryana	137.54(4)	8.74(2)
Karnataka	104.88(8)	20.04(7)
Kerala	178.68(2)	12.72(3)
Madhya Pradesh	76.79(14)	37.43(13)
Maharashtra	112.80(6)	25.02(9)
Orissa	81.00(12)	47.15(15)
Punjab	187.57(1)	6.16(1)
Rajasthan	75.86(15)	15.28(5)
Tamil Nadu	149.10(3)	21.12(8)
Uttar Pradesh	101.23(10)	31.15(11)
West Bengal	111.25(7)	27.02(10)

Source: Eleventh Finance Commission Report, HCR—Planning Commission.

While implementing various wage employment programmes government emphasises on the creation of infrastructure in the rural area. Though the basic objective of almost all the wage employment programmes is to generate income/employment opportunities for the rural poor yet in the process various rural infrastructure are created. This is inferred from the following Table 8.2.

Table 8.2: Objective of different wage-employment programmes

Sl. No.	*Programme*	*Objective in brief*	*Target area*
1.	RMP (Rural Manpower Programme)	About 2.5 million persons for 5 years by creating community assets	High unemployment
2.	CSRE (Crash Scheme for Rural Employment)	To generate 2.5 lakh man-days of employment in each of selected districts by creating productive assets and prospective irrigation	350 districts mostly SFDA, MFPA & DPAP
3.	PIREP (Pilot Intensive Rural Employment Programme)	Provide employment to 1000 persons in a year in a district and to one person in a family in which nobody was employed by creating economic and social assets	In 15 selected blocks
4.	FFW (Food for Work)	Providing gainful employment to the poor by creating rural infrastructure	In all States and Union Territories
5.	NREP (National Rural Employment Programme)	Providing gainful employment to the poor by creating rural infrastructure	In all States and Union Territories

(Table 8.2 Contd...)

Sl. No.	Programme	Objective in brief	Target area
6.	RLEGP	Provide 100 days of employment to every rural landless by creating durable assets	In all States and Union territories
7.	JRY (Jawahar Rozgar Yojana)	Generate meaningful employment opportunities for the unemployed and the underemployed in rural areas though the creation of economic infrastructure and community and social assets	In all States and Union Territories
8.	EAS (Employment Assurance Scheme)	To provide employment in the form of manual work in the lean agricultural season by creating economically productive assets.	In all States and Union Territories
9.	JGSY (Jawahar Gram Samridhi Yojana)	Creation of demand-driven community assets/infrastructure with employment generation as a secondary objective	In all States and Union Territories
10.	SGRY (Sampoorna Grameen Rozgar Yojana)	Providing additional wage employment in rural areas with the creation of durable economic infrastructure	In all States and Union Territories
11.	NREGP (National Rural Employment Guarantee Programme)	To enhance the livelihood security of people in rural areas by generating wage employment through works that develop the infrastructure base of that area.	In 200 most Backward Districts

Rural Infrastructure Development Fund (RIDF)

The Government of India set up the RIDF in 1995-96 for financing ongoing rural infrastructure projects. The fund is administered by the National Bank for Agricultural and Rural Development (NABARD). Various commercial banks contribute to the fund to the extent of their shortfall in stipulated priority sector leading to agriculture. The main objective of the fund is to provide loans to State Governments and state-owned corporations to enable them to complete ongoing rural infrastructure projects. It has been decided to extent loans to Panchayati Raj Institutions (PRIs), Non-Governmental Organisations (NGOs), Self-Help Groups (SHGs) etc., with effect from April 1999. Till the end of 2005-06, eleven trenches of RIDF (I to XI) have been completed and the XII tranche of RIDF is on progress.

The total corpus of RIDF (I to XII) amount to Rs. 60,000 crore. Cumulative sanctions and disbursements under different trenches of RIDF (I to XI) stood at Rs. 51,283 crore and 31,337 crore respectively. Out of 244651 projects sanctioned under RIDF (I to XI) 118128 projects have been completed by 31.3.2006. Table 8.3 details the matter.

Table 8.3: Trench wise RIDF utilisation (As on 31st March 2006)

RIDF tranche	*Year*	*Corpus (crore)*	*No. of states*	*No. of sanctioned project*	*No. of completed projects*	*Loan sanctioned (crore)*	*Amount disbursed (crore)*
I	95-96	2000	21	4168	4127	1906.21	1760.87
II	96-97	2500	16	8334	7570	2666.87	2397.95
III	97-98	2500	21	14346	14012	2733.82	2453.53
IV	98-99	3000	23	6172	5414	2903.32	2482.00
V	99-00	3500	25	12254	10738	3477.16	3032.66
VI	00-01	4500	25	43354	40165	4525.36	3850.83
VII	01-02	5000	23	24987	12583	4657.65	3756.82
VIII	02-03	5500	25	21012	9797	6009.36	4440.34
IX	03-04	5500	26	19605	13418	5599.18	3387.47
X	04-05	8000	25	59979	304	8289.75	2967.79
XI	05-06	8000	26	30440	0	8514.33	807.08
XII	06-07	10000					
Total		**60000**		**244651**	**118128**	**51283.01**	**31337.34**

Source: Downloaded from Internet www.nabard.

The shortfall in disbursement of RIDF as compared to sanctions continues to remain a matter of concern. The Government has taken a number of steps to address this problem.

Table 8.4 shows the benefit from RIDF. So far as irrigation facility is concerned by 31st March 2006 10.78 million hectares of potential irrigation facilities are added. During 1st tranche of RIDF the irrigation potential was created to the tune of 2.08 million hectare. However, in due course of time the effort has been slack. In RIDF XI it is raised to the extent of 1.52 million hectares. In the process 5062243 jobs were created. In RIDF V, maximum employment opportunities were created numbering 1430445. However, it was reduced from 661774 in RIDF I to 266119 under RIDF XI.

Table 8.4: Benefits out of RIDF projects (As on 31st March 2006)

RIDF Tranche	*Irrigation Sectors*		*Road and Bridge Sector*		
	Creation of irrigation potential (Million hectare)	*Recurring employment (No. of Jobs)*	*Rural Roads (length in KM)*	*No. of bridges*	*Bridge (mts.)*
I	2.08	661774	96	17	
II	0.62	220697	16619	383	24391
III	0.78	159153	18404	1381	34304
IV	0.52	132477	19286	755	40765
V	0.98	1430445	19254	756	40929
VI	0.80	787598	24026	694	37103
VII	0.65	179582	21011	3524	39530
VIII	0.78	296711	22034	844	43189
IX	1.18	499801	11360	523	33415
X	0.87	327886	25293	455	35346
XI	1.52	266119	24811		39669
Total	**10.78**	**5062243**	**202194**	**9332**	**368641**

Source: Downloaded from Internet www.nabard.

It is inferred from the Table 8.4 that a total of 202194 kms of rural road have been constructed under various RIDF. Except 1st RIDF in all other trenches an average of 20210 kms of rural roads have been constructed. So far as the construction of bridges is concerned 9332 bridges have been constructed covering a length of 368641 mtrs. It is noticed from the results that the RIDF concentrated more on construction of rural roads and bridges.

Only ongoing irrigation, flood protection, watershed management projects were financed under RIDF I as a 'last mile approach' to facilitate completion of the projects delayed on account of financial constraints. The financing of 'Rural Roads and Bridges' projects was started during RIDF II. Subsequently, coverage of RIDF was broad-based in each tranche and at present, a wide range of activities such as Primary and Secondary Schools, Primary Health Centres, Village Haats, Joint Forest Management, Terminal and Rural Market/Godowns, Rain Water Harvesting, Watershed Development, Flood Protection, Drainage, Cold Storage, Riverine, Fisheries, Fishing Harbor and Jetties, Mini/Small Hydal Projects in Power Sector, Rural Drinking Water Supply Schemes. Citizen Information Centres, Modern Abottoir, Seed/Agriculture/ Horticulture Farms, etc. are also being financed.

Bharat Nirman

To upgrade rural infrastructure, the Government of India has conceived a time bound business plan under 'Bharat Nirman'. This programme is a plan of action during 2005-09 for building up of infrastructure in rural area. A commitment of over Rs. 174000 crores has been made to Bharat Nirmal with the objective of unleashing the growth potential of our rural areas. This has six components with targets to be achieved by the end of the financial year 2008-09. These are: Irrigation, Rural Roads, Rural Housing, Rural Water Supply, Rural Electrification and Rural Telecommunication Connectivity.

Irrigation

Under the irrigation component of Bharat Nirman, the target of creation of additional irrigation potential of 10 million hectare in 4 years (2005-06 to 2008-09) is planned to be met largely through expeditious completion of identified ongoing major and medium irrigation projects.

Irrigation potential of 4.2 million hectare is planned to be created by speedy completion of ongoing major and medium irrigation projects and 2.8 million hectare is through the tapping up of ground water. The year-wise break up of target details of irrigation development under Bharat Nirman is given in Table 8.5.

Table 8.5: Overall targets of irrigation (Million Hectare)

Components	*2005-06*	*2006-07*	*2007-08*	*2008-09*	*Total*
Major and Medium Irrigation					
Completion on ongoing projects	0.90	1.10	1.10	1.10	4.20
Extension, renovation etc.	0.25	0.25	0.25	0.25	1.00
Minor Irrigation					
Surface water	0.25	0.25	0.25	0.25	1.00
Ground water	0.45	0.75	0.80	0.80	2.80
Repair, renovation etc.	0.05	0.05	0.45	0.45	1.00
Total	**1.90**	**2.40**	**2.85**	**2.85**	**10.00**

Source: Downloaded from Internet www.nabard.

Rural Roads

Government of India aims that by the end of the financial year 2008-09 every village of over 1000 population (or over 500 in hilly and tribal areas) has an all-weather road. This includes 66802 unconnected eligible villages to achieve the objective 146185 kms. of new roads is to be constructed by 2009. To ensure full farm-to-market connectivity, it is also proposed to up-grade and modernize 194132 kms. of existing rural roads. Out of a total investment of Rs. 174000 crore envisaged under Bharat Nirman investment on rural roads is estimated at Rs. 48000 crore.

The PMGSY was launched in 2001 as a 100 per cent centrally sponsored scheme with a target of connecting every habitation with a population of more than 1000 by 2004-05 and every habitation with a population of more than 500 by 2007. In PMGSY it has been decided to give one and only connection to each village. Under Bharat Nirman PMGSY has been modified to address the goal within the stipulated time frame.

The rural roads programme has been funded by the accrual through the diesel cess and external funding. Government has proposed a separate window under RIDF XII trench for rural roads with a provision of Rs. 4000.

Rural Housing

Shelter is one of the basic requirements of human survival. It assumes great significance for the rural poor as it provides an identity and sense of security to him. Hence, dispensing with the shortage of housing is an important objective in a welfare state like India. The Government of India announced a National Housing and Habitat Policy in 1998, it aims at providing 'Housing for all' and facilitates construction of 20 lakh additional houses (13 in rural areas and 7 in urban areas) per year for the benefit of the poor and deprived. To achieve this objective, a number of schemes were initiated at different points of time. These are Indira Awaas Yojana (IAY), Pradhan Mantri Gramodya Yojana (Gramin Awaas), Credit-cum-Subsidy Scheme for Rural Housing and Habitat Development, Rural Building Centres (RBC), Samagra Awaas Yojana (SAY) and National Mission for Rural Housing Habitation.

The 2001 census places the rural housing shortage figure in India at 148 lakhs. The Bharat Nirman programme has recognised and accorded due weightage to end the shelterlessness. It is envisaged to construct 60 lakh houses over next four years starting from 2005-06.

Right now the Ministry of Rural Development through the Indira Awas Yojana implements the rural housing programme. The main objective of the IAY is to help construction/up-gradation of dwelling units for the members of SC/ST, freed bonded labourers in rural areas and also for the non SC/ST rural poor living below poverty line by providing them a lump sum grant for the purpose. This is a centrally sponsored scheme whose cost is shared between the centre and state on 75:25 basis. For the allocation funds among states/UTs 75 per cent weightage is given to housing shortage and 25 per cent weightage to poverty ratio, for district level allocation 75 per cent weightage is given to housing shortage and 25 per cent of SC/ST component of the population.

Under the scheme, a maximum assistance of Rs. 25,000 per house for plain areas and Rs. 27500 per house in hilly/difficult areas is given to houseless BPL families in the rural areas. The selection of beneficiaries is done by the respective Gram Sabha and no higher approval is required. 60 per cent of the houses are to be for SC/ST beneficiaries.

An amount of Rs. 25500 core has been invested so far in IAY. About 139 lakh houses have been constructed under IAY all over the country since inception of the scheme upto January 2006.

Rural Water Supply

Drinking water and sanitation are not only the basic necessities of life but they are also crucial for achieving the goal of 'Health for all'. Nearly 80 per cent of the common diseases are caused by unsafe drinking water. Government of India along with the State Government have been trying to provide safe drinking water to rural population since the implementation of 1st Five Year Plan. However, a serious effort was made in 1972-73 through the introduction of Accelerated Rural Water Supply Programme (ARWSP) to address the problem of Rural Water Supply. The focus was physical coverage of all rural habitations with the facility of drinking water as per the following norms:

- 40 liters of safe drinking water per capita per day (lpcd) for human beings;
- 30 lpcd additional for cattle in the Desert Development Programme Area;
- One hand pump/stand post for every 250 persons;
- The water source should exist within the habitations/ within 1.6 kms. in the plains and within 100 meters elevations in the hilly area.

The scheme is a centrally sponsored scheme. Government of India provides 50 per cent of the cost and rest 50 per cent matching share is borne by the State Government/UTs. The ARWSP, however was withdrawn with the introduction of Minimum Needs Programme (MNP) during 5th Five Year Plan (1974-75). Despite the huge investment, the programme under the scheme was found to be unsatisfactory. As a result ARWSP was reintroduced in 1977-78.

Recognising the enormity and urgency of providing safe drinking water to rural areas, the ARWSP was given the shape of the National Drinking Water Mission (NDWM) 1986, which was renamed as Rajiv Gandhi National Drinking Water Mission (RGNDWM) in 1991, under RGNDWM some major reforms have been introduced to bring about sustainability in the water sector in rural India. The programme of RGNDWM may be classified into two categories: Main Programme and Supportive Activities.

The main programmes include: (a) ARWSP, (b) Sector Reform Programmes. Here the policy signified a paradigm shift from a centralized, government oriented and supply-driven programme to decentralized, people oriented and demand-driven programme.

Supportive Activities include the following:

1. Set up of human resource development cell at state level for planning, designing, implementing and monitoring;
2. Information, Education and Communication (IEC) programme aims at publicising the rural water supply programme through the mass media;
3. Set up of Management Information System (MIS) for effective planning monitoring and implementation;
4. Monitoring and investigation units have been set up at the state headquarter to carryout investigation and feasibility studies of the planned schemes;
5. Provision has been made for taking up monitoring and evaluation studies;
6. Research and Development (RD) cells have been established by the State Governments;
7. Provision has been made for drinking water in rural schools.

The goal of the Bharat Nirman is to provide safe drinking water to every habitation by 2009. To achieve the objective remaining 55067 habitations would be covered. Further all habitations, which have slipped back from full coverage to partial coverage due to failure of source and habitations, which have water quality problem (219968 habitations as per 2000 survey conducted by different State Government) to be redressed. ARWSP is the main instrument through which government tries to achieve the target. The Government of India has decided under Bharat Nirman names of habitations would have to provided by the State Government for availing funds from Government of India. This would ensure maximum transparency to the programme.

Rural Electrification

Rural electrification was taken up in India as a planned programme as early as 1951, with the ultimate objective of providing electricity to all villages. Under Bharat Nirman, Government of India has set the goal of achieving this vision target by 2009. Government of India launched the Basic Minimum Services (BMS) in 1966 to secure a better quality of life for rural people. In order to achieve the objective of sustainable human development a new initiative called Pradhan Mantri Gramodaya Yojana (PMGY) was introduced in 2000-01 bringing all the programmes covered under BMS into its ambit. The rural electrification programme has been included as a component of PMGY in 2002-03.

The Government of India has announced a new interest subsidy scheme in the budget of 2002-03 and has approved the grant of interest subsidy of over Rs. 560 crores under the Accelerated Rural Electrification Programme (AREP) during 10th Plan period. The AREP covers the electrification of un-electrified villages, hamlets/dalit bastis/tribal villages and the poor households in villages.

As per 2001 census 125000 villages are not electrified. Bharat Nirman has the target to electrify remaining villages by 2009 through the programme Rajiv Gandhi Gramin Vidyutikaran Yojana (RGGVY). Further it will also provide 23 million households with electricity.

A village will be deemed to be electrified if the following conditions are met:

1. Basic infrastructures such as distribution transformer and distribution lines are in the inhabited locality;
2. Electrification of public places like schools, panchayat office, health centres, dispensaries etc.;
3. 10 per cent of the households in a village are electrified.

Rajiv Gandhi Gramin Vidyutikaran Yojana is a Central Government scheme with 90 per cent capital subsidy to provide electrification of rural villages. Electrification of unelectrified below-poverty-line households will be financed with 100 per cent capital subsidy at Rs. 1500 per connection in all rural habitations. On 22nd May 2006, RGGVY launched a national programme of franchisees. The management of rural distribution (of electricity) will be franchisees that could be Users Association, individual entrepreneurs, cooperatives, non-governmental organisations, Panchayati Raj institutions etc.

Rural Telecommunication Connectivity

Telecom connectivity constitutes an important item of rural infrastructure. Under the Bharat Nirman programme it will be ensured that every village to be connected by telephone. The 66822 villages, which have not yet been provided, with a Village Public Telephone (VPT) shall be covered by November 2007. Out of the identified villages connectivity in 14183 remote and far-flung villages will be provided through digital satellite phone terminals. It is estimated that a total sum of Rs. 451 crores would be required for the above purpose and the entire amount will be met out of the Universal Services Obligation Fund (USOF). The government is committed to expanding rural connectivity. This will include connecting block headquarter with fibre optic network, using wireless technology to achieve last mile connectivity and operating information kiosks through a partnership of citizenship, panchayats, civil society organisations, the private sector and government.

Flagship Programme for Social Sector Development

Besides the above measures, Government of India introduced 8 flagship schemes for building up of social infrastructure from last fiscal year. These schemes are: Sarva Siksha Abhiyan, Mid-day Meal Scheme, Rajeev Gandhi Drinking Water Mission, Total Sanitation Campaign, National Rural Health Mission, Integrated Child Development Services, National Rural Employment Guarantee Scheme and Jawaharlal Nehru National Urban Renewal Mission. Except the last one the other seven flagship programmes are expected to bring about a change largely in the rural areas. The total allocation for these schemes were Rs. 34927 crore in 2005-06 and this has been stepped up to Rs. 50015 crore in 2006-07, representing an increase of Rs. 15088 crore i.e., 43.2 per cent.

Backward Region Grant Fund

The provision of rural infrastructure has been given a further boost in the Annual Plan 2006-07 by enhancing the budget provision under Backward Region Grant Fund (BRGF) to Rs. 5000 crores in the present fiscal year. It would allow the State Governments to strengthen District Programme Implementation Machinery and use the BRGF for filling critical gaps in rural infrastructure.

Conclusion

Above discussion concludes that the demand for infrastructure services continues to outpace supply. Unless these shortfalls are met by creating new infrastructure there would be serious constraints on rural development in particular and overall economic development of the nation in general. Social welfare considerations also make it imperative that a minimum supply of road, water, housing power and irrigation are made available to every citizen. Hence, the time-span in which the objective of the provision of minimum services to all can be achieved will depend on the resource allocation, political will, the administrative skill and active interest of all the stakeholders associated with the economic development of the nation.

9

India's Rural Health Services and Recent Development

*Mr. Akshaya Kumar Panigrahi**

Introduction

According to the World Health Organisation health is a "state of complete physical, mental and social well-being and not merely the absence of disease or deformity". One of the fundamental rights of every human being without distinction of race, religion, political belief, etc. is the enjoyment of the highest attainable standard of health.

The axiom 'Health is Wealth' is relevant in all ages for it is on the health of the people, that the wealth of a nation and the survival of its people depends. Where as in India large masses of the poor continue to fight a hopeless and constantly losing battle for survival and health even fifty years after India's independence. India is the second largest populated country in the world after China having 1,028,610,328 population in 2001 census in which 742,490,639 population belongs from the rural India, which constitute 72 per cent of the total population of India. Really it is a big number of comparison to other developed and some developing countries. Here Indian rural means those areas located far from towns, cities or capital towns. In other words we may say that these areas where there is no proper road connection, no drainage facilities, mass are

* **Research Scholar, Population Research Centre (PRC), Institute for Social and Economic Change (ISEC), Bangalore–72.**

poor and illiterate, depends on agriculture, lacks the minimum requirements of modern life, residing in hill sides etc. This is nothing but the life in village or gram.

Since the attainment of independence, the country has made significant progress in improving the health status of its people. Plague and smallpox have been completely eradicated, cholera has been successfully contained and considerable headway has been made in the control of malaria, leprosy, tuberculosis, blindness, filariasis and several other diseases. However, the planning process has also contributed to the development of a nationwide primary care infrastructure, reversing the kind of largely hospital based services. The main objective of this paper is to study the pattern of India's rural health services and the recent development.

While addressing the World Health Assembly in May 1981, the late Prime Minister Smt. Indira Gandhi observed:

"In India we should like health to go to homes instead of large numbers gravitating towards centralized hospitals. Services must begin where people are and where problems arise".

Health Conditions in India at the Time of Independence

In the first instance Government of India appoint the Health Survey and Development Committee known as Bhore Committee, to make a survey of the existing position in regard to health conditions and health organisation for the future development of health statistics in October 1943. The Bhore Committee which submitted their Report in 1946, found that the general death rate in "British India" was 22.4 per 1000 population, the infant mortality rate was 162 per 1000 live births, and the expectation of life at birth was 26.91 years for males and 26.56 years for females. Nearly half the total number of deaths were among children under 10 years of age and in this age group one-half of the mortality took place within the first year of life. Other health indicators are in the following Table 9.1 as shown at the time of independence.

Table 9.1: Health indicators on the eve of independence

Expected of life at birth	27
General death rate (per 1000 population)	22.4
Infant mortality rate (per 1000 live births)	162
Death rate of children under 10 years (%)	48
Maternal Mortality rate (per 1000 live births)	20
Disease wise causes of deaths (%)	
Fevers (including Malaria)	58.4
Respiratory diseases	7.6
Dysentery	4.2
Cholera	2.4
Small-pox	1.1
Plague	0.5
Others	25.8
Total	**100**

Sources: GOI (1946).

After 1946 some of the other committees are also recommended by the Government of India for studying the development of health in different time period are as follows:

1961 Mudaliar Committee

1966 Jain Committee

1974 Kartar Singh Committee

1975 Srivastava Committee

1980 ICMR and ICSSR Joint Panel

1983 National Health Policy

2001 National Health Policy

2005 National Rural Health Mission

Health Care System in India

The health care system in India consists of:

- Primary, secondary and tertiary care institutions manned by medical and paramedical personnel;

- Medical colleges and paraprofessional training institutions to train the needed manpower and give the required academic input;
- Programme managers managing ongoing programmes at central, state and district levels; and
- Health management information system consisting of a two-way system of data collection, collation, analysis and response.

Primary Health Care Services

The primary health care infrastructure provides the first level of contact between the population and health care providers. Realizing its importance in the delivery of health services, the centre, states and several government related agencies simultaneously started creating primary health care infrastructure and manpower. This has resulted in substantial amount of duplication of the infrastructure and manpower.

The government funded primary health care institutions include:

- The rural, modern medicine primary health care infrastructure created by the states consisting of: subcentres, primary health centres and community health centres as shown in chart 9.1;
- Sub divisional/taluk hospitals/speciality hospitals;
- Rural family welfare centres, urban health posts, urban family welfare centres etc.;
- Health care for Central Government employees provided by Central Government health scheme;
- Hospitals and dispensaries of railways, defence and similar large departments providing the health care to their staff.

CHART 9.1

Synoptic View of Health Care System in India

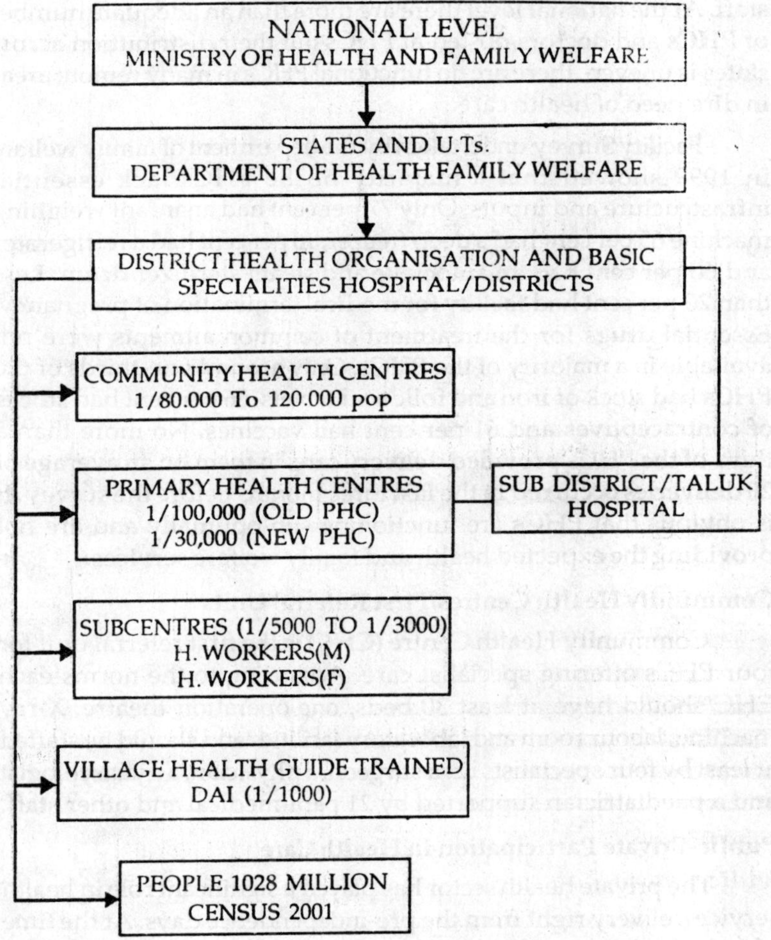

Sub-Centres

The sub-centre is the most peripheral health institution available to the rural population. Even though the sub-centre/population norm at the national level has been met, there are wide interstate variations. States with poor health indices do not have the required number of sub-centres especially in remote areas.

Primary Health Centres (PHCs)

PHC is a referral unit for six sub-centres. All PHCs provide outpatient services; a majority has four to six in-patient beds. According to the norms they have one medical and other supporting staff. At the national level there are more than an adequate number of PHCs and doctors posted at PHCs but their distribution across states is uneven, there are no functional PHCs in many remote areas in dire need of health care.

Facility Survey undertaken by the department of family welfare in 1999 showed that a majority of the PHCs lack essential infrastructure and inputs. Only 77 per cent had an infant weighing machine 65 per cent had a deep freezer, 16 per cent had a refrigerator and 60 per cent had an autoclave and steam sterilizer drum. Less than 20 per cent had facility for medical termination of pregnancy. Essential drugs for the treatment of common ailments were not available in a majority of the PHCs. Only around one-thirds of the PHCs had stock of iron and folic acid tablets, 56 per cent had stocks of contraceptives and 61 per cent had vaccines. No more than a third of the PHCs provided delivery care, in them on an average of 26 deliveries occurred in the last three months before the survey. It is obvious that PHCs are functioning sub-optimally and are not providing the expected health and family welfare services.

Community Health Centres/First Referral Units

Community Health Centre (CHC) is the first referral unit for four PHCs offering specialist care. According to the norms each CHC should have at least 30 beds, one operation theatre, X-ray machine, labour room and laboratory facilities and should be staffed at least by four specialists i.e. a surgeon, a physician, a gynaecologist and a paediatrician supported by 21 para-medical and other staff.

Public-Private Participation in Health Care

The private health sector has played a significant role in health service delivery right from the pre-independence days. At the time of independence, public private participation was in the form of government doctors being allowed private practice, an arrangement that continues even today in majority of states. To cope with the lack of medical teachers in the 1950s and 1960s many medical colleges appointed private practitioners as honorary teachers honorary physician in teaching hospital but the number of such teachers declined with the increasing availability of full time paid government teachers.

Current survey shows that the private sector health services range from those provided by large corporate hospitals, smaller hospitals/nursing homes to clinics/dispensaries run by qualified personnel and services provided by unqualified practitioners. A majority of the private sector hospitals are small establishments with 85 per cent of them having less than 25 beds with average bed strength of 10 beds. Private tertiary care institutions providing speciality and superspeciality care account for only 1 to 2 per cent of the total number of institutions while corporate hospitals constitute less than 1 per cent. There are wide inter-state differences in the distribution of private sector hospitals and beds. The private sector prefers to set up facilities in the more prosperous districts/ states especially in urban areas. The private sector accounts for 82 per cent of all outpatient visits and 52 per cent of hospitalization at the all India level with no significant variations across income group. So a majority of government and private sector hospitals and beds are located in urban areas. Qualified and registered private sector doctors or private sector institutions are not readily available in remote rural and tribal areas because people do not have ability to pay and there is a lack of social infrastructure. Thus the population in these areas where health care needs are the greatest have very poor access to functioning government health services or private facilities. In spite of the abundant supply of registered physicians in modern system of medicine and ISM & H, unqualified persons still provide health care especially to the poorer segments of the population living in urban slums, remote rural and tribal areas.

NGO and Voluntary Sector

Apart from the public and private provider of health care, the NGOs and voluntary organisation have been providing health care services to the community. It is estimated that more than 7000 voluntary organisations are involved in health care related activities. These voluntary organisations are working within a limited area with specific localities. There is also a great inter state variation among the different states in India. Some organisations are specially working on specify health related diseases like leprosy,

TB, polio, child health, women health, for older peoples health etc. Among these voluntary organisations some are working well whereas some are not working even a standard level. Some organisations are also working for the personal interest in the name of poor and ill health sections. Other than this the health care activities are also carried out by agencies like the WHO, UNICEF, Red Cross, Industrial establishments, Lion's Club, Helpage India etc. Some study shows that these voluntary organisations are facing lots of problems like non-availability of funds, limited financial support from government and international agencies, people's response etc.

India's Health Achievement Since 1951-2000

In this section we have to discuss about the demographic changes, epidemiological shifts, availability of infrastructure since 1951-2000. Under demographic changes we have to look about the life expectancy, crude birth rate (CBR), crude death rate (CDR), and infant mortality rate (IMR). From the Table 9.2, it shows that the life expectancy at birth in 1951 was 36 years, whereas in 1981 and 2000 A.D. the life expectancy became 54 years and 64.6 years respectively. It shows that within the 50 years interval the life expectancy is near to double which implies some improvements in health sector, but in comparison to China the achievement is far away. Similarly the CBR and CDR in 50s was 40.8 and 25 whereas in 2000 it became 26.1 and 8.7 respectively. More or less it is satisfactory but if we look into the natural population growth rate it became painful. In 50s the growth rate was 15.8 whereas in 2000 it became 17.4 per thousand population. But in the case of IMR it was 146 per thousand in 50s whereas in 2000 it became 70 per thousand, whereas in China and Sri Lanka it is only 41 and 18 per thousand.

Present Health Scenario in Rural India

If we look into the present scenario of health in rural India specially the death rate, infant mortality rate and age specific death rate, it is really panic in comparison to urban India (details are in the following Table 9.3). It shows that the death rate and IMR of all India, rural and urban since 1971 to 2003. No doubt within last 35 years India's death rate and IMR is decreased, but in comparison to urban India, the achievement in rural India is not good. If we look into the neo natal and post neo natal death in rural India it is really very bad. But according to age specific death rate, nearly 30 children are dying below 5 years age in rural India whereas in urban it is only 16 in 1991. There is also worse condition if we look into the IMR in comparison in rural and urban India, which is 66 and 38 in 2003 respectively. From the above analysis it shows that the urban India gains much in comparison to rural India. Similarly if we look into the interstate mortality rate it shows a lot of differences among the different states, specially in rural areas. The performance of South India is better than the northern states specially in BIMARU states.

Table 9.2: India's achievement in the years 1951-2000

Indicator	*1951*	*1981*	*2000*
Demographic Changes			
Life expectancy	36.7	54	64.6
Crude birth rate	40.8	33.9 (SRS)	26.1 (99SRS)
Crude death rate	25	12.5 (SRS)	8.7 (99SRS)
IMR	146	110	70 (99 SRS)
Epidemiological Shifts			
Malaria (cases in million)	75	2.7	2.2
Leprosy (cases perz 10,000)	38.1	57.3	3.74
Small pox (No. of cases)	>44,887	Eradicated	
Guineaworm (No. of cases)		>39,792	Eradicated
Polio		29709	265
Infrastructure			
SC/PHC/CHC	725	57,363	1,63,181 (99-RHS)
Dispensaries and Hospitals	9209	23,555	43,322 (95-96 CBHI)
Beds (Pvt. & Public)	117,198	569,495	8,70,161 (95-96 CBHI)
Doctors (Allopathy)	61,800	2,68,700	5,03,900 (98-99 MCI)
Nursing Personnel	18,054	1,43,887	7,37,000 (99-INC)

Table 9.3: Mortality estimates for all India and by rural-urban residence, 1971-2003

Period	*Crude death rate*			*Infant mortality rate*		
	Total	*Rural*	*Urban*	*Total*	*Rural*	*Urban*
1971-1975	15.5	17.1	9.8	134	144	83
1976-1980	13.8	15	8.9	124	134	71
1981-1985	12.1	13.3	7.9	104	113	64
1986-1990	10.6	11.6	7.3	91	99	59
1990-1995	9.5	10.4	6.6	76	83	50
1996-2000	8.8	9.5	6.4	71	76	45
2000-2002	8.1	8.7	6.1	63	69	40
2002-2003	8	8.7	6	60	66	38

Sources: Registrar General of India, SRS Bulletin (2005).

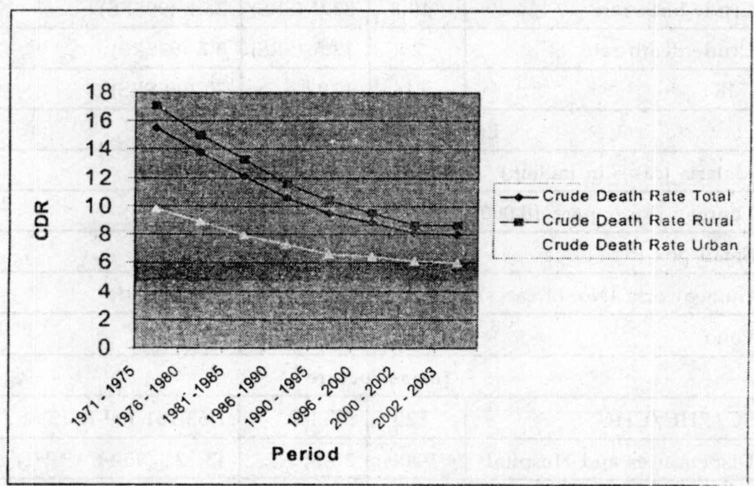

Fig. 9.1: CDR for all India and by rural-urban.

The above figure shows that the CDR is in decreasing since 1971 to 2003, but the CDR in rural India is higher than the death rate for all India and urban India.

The following figure shows the IMR for all India and by rural-urban. It also shows that the IMR is in decreasing trend, but the IMR in rural India is still higher than the urban and all India level.

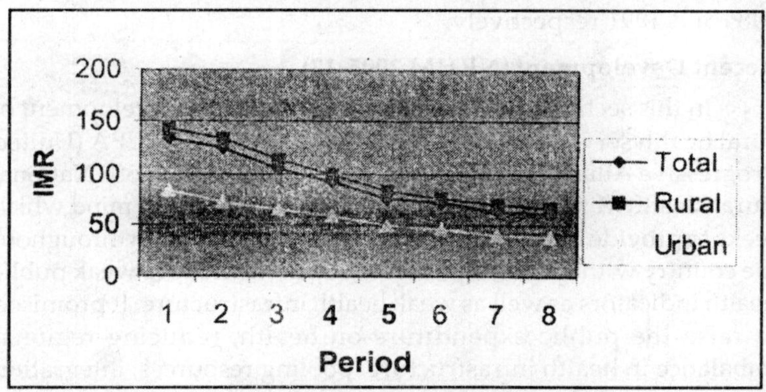

Fig. 9.2: IMR for all India and by R-U

In the following paragraph we will discuss about the main causes of death (in per cent) in rural India for two-time period, i.e. 1985 and 1991. The following Table 9.4 shows that the main causes of death in rural India.

Table 9.4: Percentage of death due to main causes in rural India, 1985-91

Group of main causes	*1985*	*1991*
Accidents and injuries	6.3	8.5
Child birth and pregnancy (Death during delivery)	1.2	1.1
Fever	9.3	7.3
Digestion defects	7.6	6.4
Cough (Breathing defects)	20.8	18.9
Defects of central nervous system	4.2	4.4
Disease of blood circulation system	9.9	11.1
Features of other clear diseases	8.9	8.3
Particular diseases of childhood	10.3	10.2
Old age	21.5	23.8
Total	**100**	**100**

Source: Jhingan, M.L. et al. 2003. Demography, p. 187.

The above table shows that the majority of persons are dying due to old age i.e. 21.5 and 23.8 in 1985 and 1991 respectively. The second highest cause is due to cough which is 20.8 and 18.9 in these two period respectively. The third highest death rate is due to some particular diseases of childhood which is 10.3 and 10.2 in 1985 and 1991 respectively.

Recent Development (NRHM 2005-12)

In this section I will discuss about the recent development of rural health services in India. On January 4, 2005 the UPA (United Progressive Alliance) cabinet approved the formation of a National Rural Health Mission (NRHM) with so many visions in mind which seeks to provide effective healthcare to rural population throughout the country with special focus on 18 states which have weak public health indicators as well as weak health infrastructure. It promised to raise the public expenditure on health, reducing regional imbalance in health infrastructure, pooling resources, integration of organisational structures, optimization of health manpower, decentralization and district management of health programmes, community participation and ownership of assets, induction of management and financial personal into functional hospitals meeting Indian public health standards in each block of the country.

The Vision of NRHM

- It seeks to provide effective healthcare to rural population throughout the country with special focus on 18 states, which have poor health indicator. It includes Arunachal Pradesh, Assam, Bihar, Chhatisgarh, Himachal Pradesh, Jharkhand, J&K, Manipur, Mizoram, Meghalaya, Madhya Pradesh, Nagaland, Orissa, Rajasthan, Sikkim, Tripura, Uttaranchal and Uttar Pradesh.
- The mission is an articulation of the commitment of the Government to raise public spending on health from 0.9 per cent of GDP to 2-3 per cent of GDP.
- It aims to involve the female activists in village level who can take care about the health of the local people in their locality, especially poor women and children.

- It seeks to revitalize local health traditions and mainstream AYUSH into the public health system.
- It aims at effective integration of health concerns with determinants of health like sanitation and hygiene, nutrition and safe drinking water through a district plan for health.

Plan of Action

For meeting the above goals the mission approves the following action as follows:

Access at the Local Level

NRHM proposes to appoint a community level health worker who has been named as ASHA (Accredited Social Health Activist). They will provide the first contact care to people in villages. These health workers will be women and will work closely with the Anganwadi workers and the ANM. They will be chosen by and be accountable to the women in the community. It implies that the involving the village women for choosing their ASHA who can provide better health for them. Definitely it is a well-developed plan for improving the rural health by involving the village people. ASHA is different from the ANM in point of that ASHA should belong to the concerned village whereas there is no such rule for appointing the ANM. In this programme the ASHA is the responsible for providing the curative assistance for fever, diarrhoea etc. and for the village concerned.

Strengthening SCs, PHCs and CHCs

The plan of action will have proposed a United Fund for local @Rs. 10,000 per annum to each village health committee for meeting the emergency health care need for the village, and supply the essential drugs, both allopathic and AYUSH to the Sub-centres. For strengthening PHCs the mission aims to provide quality drugs and equipment to PHCs and provision of 24-hour service in PHCs. But for strengthening CHCs the mission proposes the operationalizing 3222 existing CHCs as 24 hour first referral units, including posting of anaesthetists and to create new community health centres to meet the population norms as per Census 2001. Not only it proposes to strengther the health centres but also it plans to improve the rural sanitation including individual household toilets, women sanitary complex, and school sanitation programme.

Public Private Partnership (PPP)

The mission proposes to identify according to need based, thematic and geographic. Public sector to play the lead role in defining the framework and sustaining the partnership. Management plan for PPP initiatives at District/State and National level.

Role of State Government

The mission covers the entire country. But special importance on the 18 states which includes undivided BIMARU states, all north-eastern states, Himachal Pradesh and Orissa, GOI would provide funding for key components in these 18 high focus states. Other states would fund interventions like ASHA, Programme Management Unit and upgradation of SCs/PHCs/CHCs through Integrated Financial Envelop. NRHM provides broad conceptual framework. States would project operational modalities in their state action plans, to be decided in consultation with the mission steering group. It would prioritize funding for addressing, inter-state and intra-district disparities in terms of health infrastructure and indicator.

Conclusion

In conclusion I can say that after independence the Government of India have done so many programmes for the development of health to the masses. No doubt there is lot of improvement have been achieved since last fifty years of planned programme, but in comparison to the developed and some of the developing countries the performance is not good. But this time the UPA government launched NRHM in countrywide but special importance will be given to 18 states whose health are not good, are really needed. I thought it is proper step for improving the health of the masses in the grassroots level involving community participation, specially the rural people that constitutes 72 per cent of India's total population. If the programme will properly implemented then it may definitely be helpful for the development of health. Time will say what will happen in future.

REFERENCES

1. Anita, N.H., et. al., 2000. *Health and Medical Care, A People's Movement*, Foundation for Research in Community Health, Pune/Mumbai.
2. Balan, K., 1989. *Health For All By 2000 A.D.*, Ashish Publishing House, New Delhi.
3. Goel, S.L., 1980. *Health Care Administration, Levels and Aspects*, Sterling Publishers Pvt. Ltd., New Delhi.
4. ——, 2004. *Health Care Organisation and Structure*, Deep & Deep Publications Pvt. Ltd., New Delhi.
5. Government of India, 2000. *National Population Policy 2000*, New Delhi.
6 ——, 2001. *Draft of National Health Policy 2001*, MHFW, New Delhi.
7. ——, 2005. *National Rural Health Mission (2005-2012)*, Mission Document, MHFW, New Delhi.
8. India, Registrar General, 2001. *Provisional Population Totals, Census of India 2001*, Series-1, India, Paper-1, New Delhi.
9. ——, 2004. *SRS Bulletin*, Vol. 38, No. 1.
10. ——, 2005. *SRS Bulletin*, Vol. 39, No. 1.
11. Jhingan, M.L. et. al., 2003. *Demography*, Vrinda Publications Pvt. Ltd. Delhi.
12. Kamble, N.D., 1984. *Rural Health*, Ahsish Publishing House, New Delhi.
13. Kashyap, S.C., 1985. *National Health Policy*, Lok Sabha Secretariat, New Delhi.
14. Nayar, K.R., 2004. *Rural Health: Absence of Mission or Vision?* Economic and Political Weekly, Vol. 39, No. 45, p. 4872-4873.
15. Peters, David H. et. al., 2002. *Better Health Systems for India's Poor: Findings, Analysis and Options*. The World Bank, Washington, D.C.
16. Sharma, R.D., and Hardeep Chahal, 2003. *Rural Health Care Services and Patient Satisfaction*, Journal of Rural Development, Vol. 22, No. 3, p. 363-379.

10

Rural Women Entrepreneurship, Development Avenues and Constraints

*Krishna Chandra Pradhan**
*Kali Charan Pradhan***

Women constitute half the world's population, accomplish about 2/3rd of its working hours receive 1/10th of the world's income but they are considered a weaker group in economic development. However, in the context of natural development, women's participation in the economic activity is of crucial importance. The economic role played by women can not be isolated from the framework of development. In rural areas their work in the context of socio-economic settings includes a number of activities.

Empirical evidence shows that women contribute significantly to family business mostly in the form of unpaid effort and skills. The value of this effort is underestimated by the families that take it for granted. On the other hand, many of the enterprises defined as being run by women (i.e. enterprises in which women hold the controlling share) are in fact run in their names by men, who control operations and decision making. Women as a resource remains unexploited, unutilized for the betterment of their individual self and society. Entrepreneurship is an important ingredient of economic development and it is a must for the socio-economic

* **Lecturer in Economics, Khemundi College, Digapahandi, Ganjam, Orissa.**

** **An Advocate cum Social Worker.**

upliftment of rural women. With the increasing scarcity of jobs and unemployment entrepreneurship has assumed a big role and has become an important part of the economic policy of the country.

Women Enterprises

The Government of India has defined women enterprise as "an enterprise owned and controlled by women living having a minimum financial interest of 51 per cent of capital and giving at least 51 per cent of the employment generated in the enterprise to women". On this basis government offers incentives and concession to women entrepreneurs.

In India following features have been found in respect of women entrepreneurship:

- Less than 13 per cent of entrepreneurship development programme trainees are women;
- Decision relating to site selection for women are based on proximity to the home;
- Women experience greater difficulties than men in the startup stage with family problems, professional disrespect, love of technical training and marketing;
- Self assessment by women entrepreneurs shows great differences from the characteristics normally associated with male entrepreneurs, such as passive vs. energetic; non-comparative vs. comparative; private vs. social.

Rural Women and the Avenues of Entrepreneurship Development

In the past rural women concentrated on traditional activities, but now due to spread of education and favourable government policies towards self employment and skill development, women have changed their attitude and diverted towards non-traditional activities too. Thus, now a day's rural area can provide significant avenues for self employment to illiterate, skilled and unskilled women through activities in the following heads:

Handicrafts and Cottage Works

Toys, carpet making, cloth weaving, rope making, pottery are some of the examples of handicrafts and cottage works, which can be initiated and developed by the rural women to derive their income.

Leafs Paper Product

This refers to the products prepared from leaves and papers such as orts plate, use and throw food plates, dolls, stationary, items, book binding, paper bags, etc. Out of these use and throw food plates have a very potential market in the urban areas particularly in functions.

Tailoring and Dress Making

This is a profitable venture exclusively for women in rural areas. There is wide scope for tailoring and readymade garments, toys and crafts. Regional Rural Banks also provide medium term low interest loans for this initiative.

Mushroom Farming

Production procedure of mushroom is very simple and economical, moreover its demand is very high in relation to its present volume of production. Women can spare their leisure time for the production of mushrooms. By harvesting mushrooms in the rural areas they can transport it to urban areas. They can also preserve the surplus quantity in the form of pickles and mushroom nodules and sell it in the off season.

Bee Farming

Bee farming provides both gainful employment and stable source of income to rural women. This project does not require high investment and sophisticated technology. KVIC and National Honey Bee Research Centre encourage landless rural women to adopt this project. A single woman can conventionally manage upto 30 bee colonies, which can fetch her upto Rs. 1500/- per month.

Duckery and Poultry

This fetches a handsome income with in a short span of time. So poultry and duckery are gaining importance amongst the rural women.

Fishery and Dry Fishery

Fishery and dry fishery is presently being considered as viable means of employment among rural women. This sector is opening more opportunities like net weaving, mending, hand picking, seed stocking, fish breeding, processing for value added products and hygienic dry fish production, etc. In India retail fish trade is heavily dominated by women, which is the additional source of income for their family members.

Live Stock Management

Animal Husbandry and Dairy Development play a dominating role in rural economy by supplementing the additional income to rural women. It also provides ready and sure source of income to buy other inputs such as seeds, fertilizers, etc.

Nursery Raising

Fruits and vegetables has been a major food for mankind since time immemorial. They not only play significant role in human diet but well known for their medicinal values. There is a great scope of raising vegetable seedlings like tomato, brinjal, chilly, papaya and also forest seedlings. It is assumed that rural women sell the vegetable seedlings in bunches of hundred whereas fruits and forestry seedlings in single poly packs.

Floriculture

Floriculture till recently considered to be a single garden activity can be better performed by women. Besides religious offerings and home decorations now a day it is engaged as an important agribusiness enterprises. National Horticulture Board (NHB) also introduced several new schemes for women growers besides the value added products like perfumes, essential oils and medicines are promoted by Agriculture and Food Products Export Development Authority (APEDA) under the Ministry of Commerce.

Entrepreneurship Training

A number of government and non-government agencies with their varied programmes came forward to promote women entrepreneurs.

- Small entrepreneurship development institute of India (SEDII)
- District Industries Centre (DIC)
- Federation of Societies of Women Entrepreneurship
- Development of Women and Children in Rural Areas (DWCRA)

- IRDA and PMRY
- Financial assistance through Women's Development Corporations (FAWDC)
- National Backward Classes Finance and Development Corporation (NBCFDC)

To overcome the difficulties faced by women entrepreneurs in getting loans from banks and other financial institutions, some institutions have developed specialized schemes wherein finance is available to women entrepreneurs on concessional terms. Such schemes are:

Rastriya Mahila Kosh (RMK)

Rashtriya Mahila Kosh known as the National credit for women was established in 1993. It is an independent registered society for the development of women and child development in the Ministry of HRD.

Mahila Udyog Nidhi (MUN)

This is a special scheme operated through state financial corporations and scheduled commercial banks. The scheme provides self loans subject to maximum of Rs. 2.5 Lakhs per project at zero per cent interest, but one per cent service charge to meet lending expenses.

Stree Shakti Package (SSP)

This unique scheme is run by SBI to support women entrepreneurs. It is available to an enterprise where women own more than 50 per cent of the capital of the unit.

Support to Training cum Employment Programme (STEP)

The main objective of this scheme is to strengthen the skill for self employment of women below poverty line in traditional sectors of agriculture, horticulture, dairying, animal husbandry, etc. The scheme is implemented through DRDA and NGO's. The Department of Women and Child Development is the nodal agency.

Constraints Before Women Entrepreneurs

1. There might be social and cultural barriers to women involvement in a particular economic activity.

2. Women might not have a full grasp of the economic environment in which they have to operate.
3. Women might not be in a position to mobilize all resources (credit, labour, own time, etc.) required to successfully manage a specific economic activity.
4. There might not be sufficient commitment with in the community for economic activity managed by women.
5. Lake of stability especially among unmarried women. There is the chance of shifting the place of residence and ultimately collapse of the unit.
6. Non-co-operative altitude of the financial institutions towards small women entrepreneur's in rural areas.

Suggestions

The following suggestions are offered to minimize the constraints faced by women entrepreneurs:

1. Government policies on women education and their entrepreneurial development should be reevaluated and implemented in planned manners with periodical feedback;
2. To make women entrepreneurship development sustainable, a constant reinforcement is required;
3. Women should be made aware about various credit facilities, financial incentives and other allied services;
4. For effective and sustainable development of women entrepreneurship proper training focused on scientific inputs, suitable products identification, market survey, project formulation, approval procedure from the government at right time with less legal formalities and soft recovery rules are of utmost importance;
5. Self employment breeds entrepreneurship, therefore development of self employment training and programmes is primary importance;
6. As women entrepreneurs lack information with regard to their business, it is necessary to start information bureaus to help them in getting information;

7. More and more research and survey programmes should be conducted and steps should be taken to solve the problems of women entrepreneurs;
8. Frequent awareness camps should be organised by the rural development department and other nodal agencies to create awareness about different schemes of assistance available to women entrepreneurs;
9. NGO's can play a significant role in empowering women entrepreneurs by providing basic education, motivation training, financial help and so on.

REFERENCES

1. Hard Sock (1985): Money, *Sex and Power: Towards Feminist Historical Materialism*, Northern University Press, Boston.
2. UN (1995), *Report of Fourth Conference on Women*, Beijing.
3. B.V. Pillai and V.H. Kumar, *SHG's in Kerala*, Kurukhetra, Volume 54, No. 9, pp. 30-32.
4. Antony, Piush (2001), *Towards Empowerment ILO*.
5. Sharma, M, Randhawa, R, Kaur Manjit: "*Preference of Rural Women Entrepreneurs*, Social Welfare, Volume 52, No. 6, September 2005.
6. Singh Darshan: "*Strategies for Empowering Women Workers*", Social Welfare Volume 52, No. 2, May 2005.
7. Tripathy,.S.N. *Female Headed Households*, 2005. Sonali Publications, New Delhi.
8. "Tribal Development and the Provisions of PESA", Pradhan, K.C. in Dr. S.N. Tripathy's Edited Volume *Rural Development for Social Changes* (2006), Sonali Publications, New Delhi.
9. Nishi and A.K. Sah, *Women Empowerment Through Entrepreneurship*, "Employment News", Volume XXX, No. 27.
10. Dash, A.K. *Self Employment Avenues for Rural Women*.
11. Employment News, Vol. 29.
12. *Accessing Micro Credit Through RMK*, 2001, RMK, New Delhi.

11

Globalisation and Agricultural Development in Tribal Orissa

*Govinda Chandra Panda**

Introduction

For many centuries the scheduled tribes are the most neglected, deprived, excluded and disadvantaged group of people living in the bottom rung of the ladder in terms of socio-economic development. They are the missing links in the chain of economic emancipation, political modernisation and social acculturation. They have been economically subjugated, politically discriminated, socially ostracised, morally humiliated and relegated from the development main stream. Similarly, in the threshold of survival, some unsuitable government legislations have derailed the tribes either partly or completely to exercise their traditional rights over livelihood and access to resources. In spite of our serious efforts the deplorable economic environments in tribal regions have left now hundreds and thousands of the tribes impoverished, vulnerable and segregated from the socio-economic safety net. So it is quite pertinent to over-haul the existing developmental strategies and issues in relation to key economic institutions in tribal religions.

* **Mr. Panda is Head, Dept. of Economics, A.M.C.S. College, Tikabali, Kandhamal, Orissa.**

Developmental Issues and Key Economic Institutions: Some Realities

Since independence, crores of rupees have been invested and plenty of experiments have been made still basic contradictions which were haunting the causes of law sustainability, adaptability and productivity of key economic institutions of tribal India have continued to prevail even today. Forest and Agriculture are supposed to be the key institutions in determining the socio-economic environment of the tribal regions.

(i) *Forest*— To Prof. Ghurye, the tribes are the forest dwellers and the original inhabitations of the soil (Adivasi). This clearly speaks that the relationship between the tribes and the forest is historically inseparable. Forest has been playing a vital role in providing livelihood to tribes for many centuries. But, high consumerism and excessive commercial use of forest (instigated by non-tribes) now have resulted in first shrinking of forest resources in tribal regions Deforestation for the purpose of industrialisation has also brought endless losses to the tribes and the forest in terms of flora, fauna and collection of minor forest products. Displacement for big irrigation projects also puts the tribes in fix to exercise their legal rights over the forest lands because of non-holding of Record of Rights (RoR). Now it seems, the age-old emotions of the tribes towards the forest are gradually ending in a sorry note and they are becoming foreigners' in their own forest land. To conclude forest is no longer serving as a big and sustainable economic zone for the tribes in the field of income, employment and output.

Agriculture

Agriculture is the main stay of the tribal economy. It along with the allied services play a significant role in terms of its percentage contribution to household income, capacity of generating additional output and potentiality of creating employment opportunities in tribal regions. Further, especially with the emergence of the New Economic Policy of Globalisation (discussed in part-II) in early 90's and with the adoption of the 'Second Phase of Green Revolution' the agrarian economy of the tribes has been supposed to undergo a complete technical transformation. Agriculture in tribal regions is expected to:

(i) Promote sectoral diversification

(ii) Widen the domestic market through higher purchasing power

(iii) Facilitate inter-sectoral transfer of capital

(iv) Develop a balanced infrastructure for social capital formation and human resource development

(v) Ensure better quality of life through assured food security.

But, in spite of these high economic values, the major concerns of agriculture in tribal regions are its multi-factorial problems like depletion of ground water, deterioration of forest resources, degradation of soil, health, soil erosion, ecological imbalance, water logging, uneven rain, increase in soil salinity and decrease in cultivable land areas which finally make agriculture unpredictable and risk-prone in tribal regions. Other geographical obstacle which pose serious threats to tribal agriculture are the unsuitability nature of the lands for cropping pattern because of high mountain terrains, plateaus, and slopes, globalisation, is now putting some new challenges to already sick agriculture in tribal regions. So, it seems imperative to re-examine agricultural strategies in tribal regions so as to transfer it from 'self-deficient' to 'self-sufficient' and 'sustainable' in the new era of globalisation.

Objectives of the Study

In this back drop, attempts have been made in this paper to explore the following objectives:

1. To study the tribal agrarian economy vis-à-vis globalisation;
2. To examine credits to agriculture and the pattern of expenditure on the basis of the socio-economic status of the respondents;
3. To analyse the production in the agricultural sector and availability of marketing facilities;
4. To suggest the policy majors further policy paradigms.

Methodology

Keeping in view of the objectives of the study, the district of Kandhamal has been selected purposively due to its high tribal concentration and dominance of agriculture in the economic life of the tribes. But the finding of the study can also be generalised to other parts of the country with the similar background. For this purpose, helps have been taken from secondary sources such as journals, magazines, reports and office files to study the credits supplied to agriculture, agro-environment and the trends of production in the agricultural sector. To examine the pattern of expenditure and availability of marketing facilities data have been collected from the primary sources.

Analysis

This section intends to present the entire study in three parts. Part I exhibits the agricultural profile of Kandhamal District including the trends of productions and pattern of expenditure of credits supplied on the basis of observations and opinions collected from the farmers. Part II is the core of this research write-up. It aims to raise some fundamental issues relating to globalisation and agriculture in a tribal district with reference to marketing and productivity. Part III highlights the suggestions and concluding remarks.

PART-I

Kandhamal Agricultural District—A Brief Profile

Part I of the study attempts to highlights a brief profile of Kandhamal Agricultural District along with the trends of production of Rabi and Kharif crops. It also reveals credits supplied to agriculture and the pattern of expenditures.

A. The Agro-Climatic Profile

- ❖ It is a centrally located district and lies in between 80°30′ to 84°35′. East Longitudes and 19°34′ to 20°34′ North Latitudes.
- ❖ It is situated at an elevation of 300 MTs. to 1100 MTs, just in the middle of the State of Orissa under the North-Eastern Ghat Agro Climatic Zone.
- ❖ The soil of the district is red laterite group (Red sandy). It is light textured and acidic in nature having PH. 5.3 to 6.3. The soil is very porous with low moisture holding

capacity subject to heavy run off and soil erosion during Kharif season.

- The rainfall in uneven in nature. The average rainfall is 1630 mm in 85 rain days.
- The temperature of the district also varies from 0° Celsius to 45° Celsius. The climate is sub-tropical in nature characterised by hot and dry in summer, cool and humid monsoon, cold and dry in winter.

B. The Administrative Profile

- Deputy Director of Agriculture – 1 No.
- District Agriculture Officer – 1 No.
- Additional District Agriculture Officer – 3 Nos.
- Junior and Assistant Agriculture Officer – 24 Nos.
- Village Agriculture Workers/ Agriculture Officer – 158 Nos.
- Seed Sales Centres – 24 Nos.
- Commercial Banks Financing Agriculture – 31 Nos.
- Cooperative Banks Financing Agriculture – 6 Nos.

C Classification and Pattern of Use of Lands

- Total Geographical Area – 7649 Sq. K.M.
- Forest Coverage – 3624 Sq. K.M.
- Cultivated Land – 127790 Ha.
    - High Land – 96298 Ha.
    - Medium Land – 20624 Ha.
    - Low Land – 10868 Ha.
- Paddy Land – 57000 Ha.
    - High Land – 25517 Ha.
    - Medium Land – 20624 Ha.
    - Low Land – 10859 Ha

- ❖ Irrigation facilities available – 15334 Ha. (12%)
  - Khariff – 10223 Ha. (8%)
  - Rabi – 5111 Ha. (4%)
- ❖ Cultivators – 102929
  - Marginal farmers – 54680
  - Small farmers – 26717
  - Big farmers – 21532
- ❖ Agricultural Labourers – 71170
- ❖ Fertilisers used (N+P+K) – 319.911 MT
  - Rabi – 104.163 (3.437 Kg./Ha.)
  - Kharif – 215.748 (1.869 Kg./Ha.)

D. Credit Supplied by Commercial and Cooperative Banks/ Societies

- ❖ Cooperative Banks
  - ***Sanctioned*** – 365.30 Lakh
  - Cash – 68.50 Lakh
  - Kind – 296.82 Lakh
  - ***Disbursed*** – 120.87 Lakh
  - Cash – 35.32 Lakh
  - Kind – 85.55 Lakh
- ❖ Commercial Banks
  - Sanctioned – 4.50 Lakhs
  - Disbursed – 2.28 Lakhs

E. Trends of Production and Pattern of Expenditure

Trends of production have been depicted in Annexure-I for Kharif and Annexure-II for Rabi Crops. Similarly the pattern of expenditure of credits has been discussed in the summary part of this section.

F. Summary of Part-I

An overview of this section clearly reveals that agriculture is affected by multiplicity of problems in the tribal district of Kandhamal. Forest covers nearly 50 per cent of the total geographical land area and leaves less lands for cultivation. Agriculture thus becomes insufficient to adjust the growing labour force. Extreme climates also affect in production of certain crops like jute, sugarcane and spices. Similarly, uneven rain causes regular incidents of floods and droughts. Crop failure and lack of insurance put the tribes in indebtedness. The become easy victims of poverty and exploitation by the moneylenders. Land classification also shows the presence of high lands in more quantity causing unsuitability in the production of major crop like rice. Inadequate irrigation and supply of power lead to low productivity in agriculture. Less use of chemical fertilisers and pesticides keep the tribal farmers miles backward to produce crops in a tune with the average yielding per hector. The situations of market are also not in favour of protecting the economic interest of tribes by providing the minimum support price for their products. Illiteracy cause high concerns in tribal regions to study the conditions of price mechanism and price structure by the tribes. Irregular supply of credits force the tribes to seek financial supports from the business houses and non-financial institutions. Credits availed by the small farmers are mostly used for non-agricultural purposes. Again less amount of credits are utilised for the purpose of infrastructure development. Big farmers often use their credit in lending loans to the small farmers. Ironically the commercial banks are playing a minor role in providing financial support to agriculture in tribal Kandhamal. There is also disproportionate in supply of credits in terms of cash and kind. The cooperative banks supply credits more in kind which lead to less capital investment of agriculture.

PART-II

Globalisation and Agriculture—Needs Some Strategic Discussion

The economic reforms of early 90's in the form of globalisation have changed the very dimensions of Indian economy. Growth guided by market forces and polices of reforms seems to be the guiding principle. Globalisation has brought a competitive market and players with high level of technology and cost-effective techniques have emerged in the economy. It promotes the culture of consumerism and ultimately helps to generate wealth and employment all over the world. It rests on the philosophy of 'market access' and calls for a "no-intervention regime". The whole doctrine is based on the principle of laissez faire and the market ethics is competition...Competition and more competition. Methodologically, the feel-good factors of globalisation are in terms of:

(i) Increase of world income, out-put and employment via national economies;

(ii) High consumerism and surpluses to the consumer in the way of price-utility at the consumption end;

(iii) World technology innovation and change in the scale of production;

(iv) Least-cost combination in production resulting in low level of price;

(v) Mobility of capital and labour, thereby bringing equilibrium in the factor market;

(vi) Reduction in poverty, unemployment in inequality;

(vii) Improvement in the standard of living by the way of producing quality outputs;

(viii) Minimising the market deficiencies to ensure better support prices for industrial and non-industrial, agricultural and non-agricultural, consumer's and producer's goods.

Now, the fundamental question is, which sector of the economy will survive in this harsh globalised economic environment? The answer is definitely not the agricultural sector of the tribal economy. Gradually the economy is paying heavy social cost in terms of agriculture. Globalisation ensures increased output employment and income. But, the size of land holdings, the land distribution pattern, yield per hectare and land classifications clearly indicate that agriculture is still miles backward in providing even the basic necessities of human life. Most of the studies reveal that poverty remains yet unabated in tribal regions because of serious decline in agricultural activities. Cut in subsidies and insufficient release of credits also result in the squeeze of employment opportunities in the tribal region. Uneven rain and less use of fertilisers make agriculture inefficient to be a big player in the field of occupational mobility and structural diversification. The agricultural scenario in tribal hinterland is characterised by low rate of tenancy reforms, landlessness, high degree of fragmentation and skewed distribution of holdings. Similarly lack of irrigation facilities and improper management of water stock cause heavy concerns to tribal agriculture in the field of generating additional employment opportunities for the growing labour force. The pressure of population on land and the features of under employment, seasonal unemployment and disguised unemployment make agriculture over-burden in the tribal regions. The geographical situations also turn agriculture negatively in providing sufficient food to the growing millions in a sustainable manner, year after year, without compromising the future capacity of our natural resources. Widespread deforestation aggravates the situations further and brings imbalances in the eco-system by the way of uneven rain. These fundamental problems keep agriculture in dormant position to participate in the whole process of globalisation. These situations ultimately lead to worst standard of living by the tribes in terms of human resource development. It creates "a reserve army of unemployed agricultural workers" with high propensity to consume and low purchasing power. High propensity to consume leads to low capital formation, low saving and low investment causing serious shortage of capital needed for investment. So, it naturally becomes a Herculean task especially for the small and marginal farmers to participate in the market process with positive outlooks. The big agricultural houses play a upper hand in determining the prices of agricultural products. Neither the invisible hand of price mechanism nor the forces of demand and supply help the small farmers to get the minimum support price. The agrarian market structure also helps the outside producers in many ways and becomes conducive for them to interfere in the process of both commodity and factor pricing. It becomes easy for them because they always enter in the market with low priced commodities produced basically with the application of new technologies at a lower scale of production. Thus, the new era of globalisation instead of bringing agricultural prosperity ends up in agricultural disaster by leaving most of the tribes in indebtedness. It rather helps the emerging class of progressive cultivators "The nouveau riche" of tribal India. Insufficient financing also forces the tribes to seek financial support from the village moneylenders at a higher rate of interest. They usually take loans before cultivation with an unwritten agreement to sell the their products in the later part. As a result, they are unfortunately given thrown away prices much below the market rates in the whole process. Sometimes, the production costs even exceed the sales price. What happens ultimately, the economic powers polarise in few hands. Polarisation of powers result in the outflow of capital from tribal to non-tribal hands. Globalisation instead of ensuring mobility of labour and capital promotes drainage of capital by causing heavy damages to the agricultural base of the tribal region. Illiteracy is another stumbling block which prevents the tribes to make agriculture technology efficient. Application of information technology is hardly found in the field of crop diversification, extension services, technological know-how and adoption of high yielding variety seeds. Sectoral disequilibrium is another major concern which leads to unbalanced agricultural growth because of strategic gap between the industrial and agricultural sector. In the conclusion it can be said that agriculture is already in the cross road as globalisation has changed the production base from socialisation to commercialisation but it is yet viewed from a societal point in tribal regions.

PART-III

Suggestions and Concluding Remarks

Suggestions

The following suggestions put forth can make agriculture in a tribal region sustainable in the new era of globalisation:

(i) Increasing use of specific high yielding crops and short duration crops in specific location;

(ii) Building awareness and confidence among the farmers for crop diversification from paddy to more remunerative non-paddy crops like pulses, oilseeds, vegetables and other horticultural crops, sugarcane, cotton;

(*iii*) Encouraging cultivation of pulses and oilseeds as mixed cropping in the rain fed farming system to avoid crop failure in drought like situation;

(*iv*) Timely pre-positioning of the inputs like seeds, fertilizer, pesticide, implements and crop loan in adequate quantities;

(*v*) Encouraging farm mechanization by supply of tractors, power tillers, power operated implements and self propelled implements etc.;

(*vi*) Increasing use of fertilizer, emphasising on their balance used (N+P+K) with popular blending or micronutrients (INM) along with application of organic manures, bio-fertilisers;

(*vii*) Creating awareness among the farmers for soil testing;

(*viii*) Propagating the adoption of integrated pest management approach for effective control of pests, insects and plant diseases;

(*ix*) Expansion of irrigation facilities through private L.I. points, dug well and river lift points;

(*x*) Bringing the latest crop production technologies to the doorsteps of the farmers through effective extension and educating the farmers for adoption;

(*xi*) To encourage farmers to avail credit as per requirement;

(*xii*) Growing of high value crops like vegetables and spices and transfer of technologies for increasing production and productivity.

Concluding Remarks

A perusal of the study clearly acknowledges that agriculture in tribal regions is in low profile particularly in the context of globalisation and there is urgent need to rectify the situation. The economic reforms of early 90's in the form of globalisation has opened up new challenges, threats, prospects, opportunities and potentialities in the agricultural sector of the Indian economy. Accordingly, the entire agrarian sector has been influenced by the situation of progression and retrogression, acceleration and deceleration.

As far as tribal agriculture is concerned globalisation totally fails to meet the hopes and aspirations of tribal cultivators in providing an economic safety net. Globalisation can hardly bring any change in the economic life of the tribes via agriculture because their economy is still packed with some fundamental problems like unconquered poverty, unwieldy unawareness, inaccessible illiteracy, uncontrollable population growth, dreaded social beliefs, unparallel disequilibrium and irremovable inequalities. Unless these problems are tackled, globalisation will undoubtedly bring economic hardship to tribal agriculture and may lead to dehumanisation of workforce and unfair competition in agricultural market. Again the shift from subsidy economy to market economy has also brought on sustainable loss to social sector development including the agriculture and allied services in the tribal regions. Thus in the matters of agricultural policy formulation; its growth can not be left entirely to market mechanism. The market forces have their own limitations and cannot also determine the development strategies alone. Hence, agriculture needs some protective approach particularly in the field of credit, market, subsidy and infrastructure to make it integrative and sustainable. To conclude, the feel good factors of globalisation are looking methodologically brighter, but the agricultural sector of the tribal economy in reality is not shining at all.

ANNEXURE-I

Area Productivity and Production Trend of Different Crops in Kandhamal District (Kharif)

A- Area in Hectare Y- Yield Rate in Kg/HA P- Production in M.T.

Sl. No.	Crop Name	2001			2002			2003		
		Area	*Yield*	*Production*	*Area*	*Yield*	*Prodn*	*Area*	*Yield*	*Production*
1	2	3	4	5	6	7	8	9	10	11
1.	Local Paddy	16830	1982	33359	8898	1212	10782	9159	2160	19783
2.	HYV Paddy	40041	2437	97583	38322	1570	60146	42071	2805	118850
	Total Paddy	56871	2300	130912	47220	1502	70928	51230	2690	137833
3.	Local Maize	7358	1178	8671	5914	1152	6872	6012	1140	6864
4.	HYV Maize	6785	1655	11223	6998	1610	11275	7682	1615	12410
	Total Maize	14143	1406	19899	12912	1405	18147	13694	1407	19274
5.	Ragi	2241	510	1143	2287	495	1133	2005	535	1073
6.	Jowar	528	415	219	547	409	224	461	395	182
7.	Bazara	180	350	63	210	315	66	175	320	56
8.	S. Millet	1680	220	358	1104	225	248	538	215	116
	Total Cereal	75571	2020	152624	64280	1412	90746	68103	2325	158534
9.	Mung	1136	197	224	758	196	149	858	227	195

(Annexure-1 Contd...)

1	2	3	4	5	6	7	8	9	10	11
10.	Biri	4793	290	1103	3498	235	820	3730	243	908
11.	Arhar	5252	828	4351	5137	848	4361	4908	885	4352
12.	Com Pea	3797	390	1479	1485	368	547	1041	375	389
	Total Pulses	14978	477	7157	10878	540	5877	10537	554	5844
	Total Foodgrain	90649	1764	159781	75158	1285	96623	78640	2090	164378
13.	Gound Nut	1050	955	1003	1270	966	1228	1420	1555	2211
14.	Niger	11550	195	2256	7756	165	1280	10619	215	2280
15.	Til	435	183	80	696	180	125	634	197	125
16.	Caster	414	760	315	382	814	311	354	780	277
17.	Sun Flower	31	600	19	11	600	6.6	72	583	42
	Total Oil Seed	13840	272	3673	10115	292	2950.6	13099	375	4935
18.	Nesta	140	345	48	112	357	40	120	375	45
19.	Sun Hemp	130	382	51	63	380	24	76	395	30
20.	Cotton	217	294	64	74	419	31	50	600	30
	Total Fibre	487	384	163	249	380	95	246	425	105
	Total Agril. Crops	104516			85522			91985		

(Annexure-1 Contd...)

1	2	3	4	5	6	7	8	9	10	11
21.	S. Patato	1578	5850	8443	922	5458	5033	1068		
22.	Other Vegetable	16711	9146	152851	12878	9238	118967	13850	5452	5823
	Total Vegetable		8820	161294	13800	8985	124000	14918	9263	128298
23.	Turmeric	9815	4860	47699	12222	5019	61351	12066	8990	134121
24.	Zinger	2835	5518	15645	3318	5691	18884	2961	5038	60790
25.	Chilly	863	508	439	565	485	274	666	5675	16806
	Total species	13513	4720	63783	16105	5000	80509	15693	495	331
	Total Hort Crop	31802			29905			30611	4965	77927
	Grand Total	136318			115427			122596		

ANNEXURE-II

Area, Productivity and Production Trend of Different Crops in Kandhamal District (Rabi)

A- Area in Hectare Y- Yield Rate in Kg/Ha P- Production in M.T.

Sl. No.	Crop Name	2001			2002			2003		
1	2	3	4	5	6	7	8	9	10	11
1.	Paddy	333	2828	942	199	2944	586	166	3138	521
2.	Wheat	84	815	68	78	885	69	120	908	109
3.	Maize	97	1250	121	76	1290	98	74	1325	98.4
4.	Ragi	17	750	13	5	600	3	5	520	2.6
	Total Cereals	581	2154	1144	358	2085	756	365	2002	731
5.	Gram	80	340	10	61	310	19	43	302	13
6.	Muga	56	170	10	33	182	6	44	181	8
7.	Biri	131	225	29	69	245	17	66	212	14
8.	Kulthi	10443	175	1827	12617	178	2252	13725	177	2428
9.	F. Pea	251	370	93	333	380	127	394	406	160
	Total Pulses	10911	180	1969	13113	185	2411	14272	184	2623
10.	S. Flower	63	470	30	30	466	14	23	521	12

(Annexure-2 Contd...)

1	2	3	4	5	6	7	8	9	10	11
11.	Safflower	3	195	0.6	3	133	0.4			
12.	Mustard	13708	165	2262	10415	166	1730	14060	170	2404
13.	Til	171	120	20	466	137	64	238	139	33
14.	G. Nut	14	1150	16				16	1000	16
	Total Oilseed	13959	167	2328.6	10914	165	1808.4	14339	172	2465
15.	Onion	138	4205	580	139	4170	580	146	4267	623
16.	Patato	1326	1256	6969	1255	4635	5817	1216	4697	5712
17.	Other Vegetable	2476	9585	23732	4100	9512	39000	2850	9507	27095
	Total Vegetable	3940	7939	31281	5494	8263	45397	4212	7936	33430
18.	Chilly	184	370	68	187	358	67	185	367	68
19.	Garlic	1981	1965	159	87	1955	170	91	1978	180
20.	Corriander	51	295	15	57	315	18	57	298	17
	Total Spiecs	316	765	242	331	770	255	333	795	265
21.	Tobacco	72	330	24	86	313	27	81	333	27
22.	Sugarcane	4	13460	54	3	48000	144	5	41600	208
	Total	29733			30299			33607		

REFERENCES

1. Tripathy, S.N. "Education for Tribal Women" in Self Edited *"Tribal Women in India"*, Mohit Publication, New Delhi, 2002.
2. Panda, G.C. and Tripathy, Dr. S. "Impact of Road Transport on Population: A Study in Kandhamal District, Orissa." in *"Problems of Population in India"* Edited by Panigrahy, R.L., Discovery Publishing House, New Delhi, 2004. (In Press).
3. *Statistical Outlines of Orissa—2001*: Directorate of Economics and Statistics, Govt. of Orissa, Bhubaneswar.
4. *Smaranika Kandhamala Mohouchhaba 2003*: District Council of Culture, Kandhamal District, Phulabani.
5. *District Statistical Handbook Kandhamal—1999*: Directorate of Economics and Statistics, Govt. of Orissa, Bhubaneswar.
6. Panigrahy, R.L., *Human Resource Development and Labour Welfare*, Mohit Publication, New Delhi, 2002.
7. Panda, G.C. *Panchayati Raj Institutions and Tribal Development*, Paper Presented in the National Seminar (January 24-25, 2005) at School of Social Sciences, SRTM University, Vishnupuri, Maharashtra.
8. Dhar, P.K., *Indian Economy*, Kalyani Publishers, Ludhiana (1999).
9. *Agenda Note on District Agricultural Strategy Committee Meeting of Kandhamal District*, Dept. of Agriculture, Govt. of Orissa.
10. Sangwan, S.S., *Diversification in Agriculture—How To Do It*? Kurukshetra, Vol. 50, No. 4, February 2002.
11. Agarwal, N.L., *Agricultural Marketing*, Yojana, Vol. 48, October 2004.
12. G.S. Ghurye, *"The Scheduled Tribes"*, 2nd Edition, Bombay' 1959.

12

Agricultural Marketing and Empowerment of Farmers in Tribal Orissa

*Dr. B. Eswar Rao Patnaik**

SECTION-I

Introduction

J.S. Uppal defines "marketing refers to activities involved in the exchange of commodities from producers to consumers and all physical commodities designed to give the commodities increasing time, place and form utility. The Eight Five Year Plan at country level has mooted on accelerating exports of agricultural commodities from India, by assigning the status of industry to agriculture. The major planks of agricultural policy, 1996 of the Government of Orissa are doubling the production of foodgrains and oil seeds by the end of Ninth Plan, and encouragement to private enterprise in the field of agriculture, by welcoming private partnership in marketing of agricultural products.

The economic reforms initiated at country level in 1991 did not touch agriculture. The nation has to pay heavy price for the negligence of agriculture in terms of suicidal deaths of several farmers in the States of Andhra Pradesh, Maharashtra and Orissa (Bargarh District). Dismal failure of cotton crop is the villain behind the scene. The recent enquiry into farmers deaths in Vidarbha

* **Reader and Head of Economics, SBRG Women's College, Berhampur, Orissa.**

District of Maharashtra by M.S. Swaminathan, the Chairman of National Commission on farmers may not cut much ice, because, freedom loving farmers of Vidarbha have been cotton growers, since time immemorial. Measures like, rescheduling of loans to distressed farmers and grant of lakhs of rupees to the relatives of farmers are partial answer to the crisis.

Of late, India, has emerged as a powerful market globally in cotton, thanks to the use of G.M. seeds. As Prof. Yoginder K. Alagh observer (New Indian Express, 2006, September 30) we have to build up the competitive strength of the farmers by empowering him by measures like, increasing the income flowing to him. There is nothing wrong with market based reforms. You have to adopt it to a country called India and integrate it with peasants. Efforts may be made to build up capabilities of farmers to negotiate for his land. Cooperatives and corporate sector would come to terms with market India's greatest competitive strength the labour advantages of its agriculture. The recently concluded Chief Ministers conference of Congress led governments under the leadership of Sonia Gandhi and Prime Minister Manmohan Singh at Nainatal has deplored at stagnation in farm production at Country level and their policy thrust was on bringing additional 1 crore land under irrigation and decentralization of Panchayat system.

Methodology

The present study is an end product of investigation carried out by the scholar in Koraput district in 1987 involving a total of 389 households. The academic exercise is based on Primary data, supplemented by secondary data, assembled from Economic survey 2003-04, Government of Orissa, Dailies like, Indian Express, Census of 2001, Orissa Development Report, 2002, thesis of the scholar and edited works published by Dr. S.N. Tripathy. Scanty research works have been made on marketing aspect of Koraput District. So, the present study is taken up the findings of which may have relevance at policy making level.

SECTION-II

Definition of Agriculture Marketing

Marketing is a process, whereby the producer and the buyer are brought together (A.N. Agarwal). Agriculture as an enterprise is neither a single man nor a single stage operation. First, the product must be collected from farmer's premises and assembled in bulk. The heterogeneous output produced is sorted out into grades for bringing out a co-ordination between producers and consumers. Thirdly, the raw materials produced on farms may require processing, storage for ensuring regular supply of food stuffs around the year. Transport from place of assembly to that of final sale is an urgent necessity. The provision of credit and bearing of risks are other partners in the drama of agricultural marketing.

Types of Market Organisation

J.S. Uppal reasons that, a well developed market may inject more money into the body of the economy, stimulate demand and thereby widen the employment base of the economy. The Indian economy was served by 7060 regulated markets in 1998. There are 106 weekly and bi-weekly markets locally known as "Hats" in the agriculture based district of Koraput and they serve an area of 60 miles maximum. Altogether, there were 22,000 primary markets in the labour surplus economy of India.

The industrially backward district Koraput has 5 whole sale markets, known as "Mundies", below the state average of 76 and national average of 6,000. The wholesale markets are owned by private persons or municipalities, with representatives from government legal bodies traders, agents and consumers. These markets fix commission to be charged, hear complaints, settle disputes and licenses brokers and middle man.

Sales Operations in Study Areas

The study has flashed that, sales operations to the tune of Rs. 8,46,733 in Koraput have involved 119 (40%) of households. Bulk of the sales are affected within the village to it itinerant traders. The present producers, borrow from usurious crafty moneylenders in form of paddy and return double the amount of produce after harvest.

Holding-wise Sales

One dimension of the study is related to increase in the quantum of sales with an increase in the size of land holding. Out of total sales of Rs. 12,48,57.00 in study areas in 1987, the average sales of marginal farmers and small farmers was Rs. 1363.68 and

3298.4 respectively. At the other end of the scale volume of sales by semi-medium, medium and large farmers was Rs. 7383.0, 10114.0 and Rs. 41,618 respectively. The line of reasoning is that, smaller farmers have small waiting capacity due to which they sell at a time, when price trends in the market are not favourable.

Infrastructural Facilities

While, Jeypore region is well served by F.C.I. such facilities are conspicuous by their absence in the backward block of Narayanapatna. Parallel developments in Orissa, reveal the completion of 5 godowns against the sanctioned number of 7 godowns, with a capacity of 2000 million tones at an expense of 154.30 lakhs by the centre.

Even more revealing, is the inadequacy of available infrastructural facilities in the region. The backward district has 72.00 km. of roads for 10000 population, which lags behind the state average of 140.41 km. Further, only 1 per cent of roads are maintained by N.H.S. and merely 6 per cent of villages have metalled roads.

Sales of paddy, wheat and cattle may be direct or indirect. In indirect sales, middle men operate to establish a link between producer and seller.

SECTION-III

Public Distribution System

The public distribution system which had its origin at the recommendations of Dharia committee is expected to counterweigh price rise on the one hand and on the other hand it promotes distributive justice by supplying essential commodities like rice, wheat, sugar, soft cake, kerosene and controlled cloth. In Orissa Civil Supply Corporation (OCSC) looks after distribution of essential commodities, which are allotted by Government of India. The above essential commodities were distributed among 81 lakhs ration (Card) holders. The district's economy is served by 1988 fair price shops below the state average of 24878.

Organisational Set Up

It is well known that, both at country level and state level Gram Panchayats, cooperatives LAMPS, Consumers, cooperative stores and Regional Marketing co-operative societies deal in sales and purchase operations in place of private traders. LAMPS took the initiative of purchase and sales of hill brooms, Kusumo seeds, Siali leaves and tamarinds in tribal areas.

Ever since, 1/10/1997, the centre and the states have been concerned with distribution system, also known as targeted public distribution system in I.T.D.P. and D.P.A.P. blocks.

The public distribution system, may reduce poverty in the economy and ensure food security. People with an annual income not exceeding Rs. 11000 in these blocks are supplied with 16 kg. of common rice per family at Rs. 4.75 per kg. per family per month. Further, BPL families listed under DRDA are supplied with 25 kgs. of common rice per family at a subsidised rate of Rs. 6.30 per kg. during 2003-04.

Defects of the System

Dr. S.N. Tripathy discerns that, the Government institutions have been able to procure only 10 per cent of market arrivals of minor forest produce and surplus of agricultural products. There were fluctuations in collection of tribal products by government. As per the observations of Orissa Development Reports 2002, at all India level rice purchased from PDS form nearly 16.7 per cent of total rice purchased by people. Kirit S. Parikh was pained to note that, in states like, U.P., Bihar and Orissa 98 per cent of rural population do not make any purchase from Public Distribution System. It follows that, a microscopic minority of rich have been benefited by PDS.

In PDS, the quantity of different commodities supplied to poor are not sufficient to maintain their livelihood and physical accessibility of poor to fair price shops is less as shops are located far away from houses of farmers. All essential commodities required by farmers are not supplied through PDS.

Cases of black marketing in essential commodities is not novel in Orissa. So, with a view to giving effect to Black marketing and maintenance of essential commodities Act 1980, the civil supplies department has made 73279 checks in 2003 and following 491 raids 879 cases were detected.

SECTION-IV

Procurement Price and Income of Farmers

Academicians, like Uma Lele contend that, an increase in agricultural prices may induce progressive farmers to assume risks in the adoption of new techniques with saving pool. Incentive prices are of pivotal importance for the economy, when many markets tend to be glutted during harvest season leading to distress sales, in village economies. The purchase of cereals at procurement price may counter a possible decline in price of food products. The late Rajkrishna has reasoned that, price guarantee may influence decision in production of specific crops and it may induce cultivators to use traditional and new inputs.

An increase in the prices of farm inputs used by cultivators seems to be an explanation for upward revision of procurement prices. Thus, the irrigation rates in Orissa comes to Rs. 16/- Rs. 12/- and Rs. 8/- per an acre of 1st class 2nd class and 3rd class land. Dr. Baidyanath Mishra, the former Deputy Chairman of Planning Commission in Orissa argues that, the support price of wheat has increased by 2.26 times between 1970-71 to 1988-89 from Rs. 76/- per quintal of wheat in 1970 to Rs. 178/- in 1988-98 but the prices of non-agricultural commodities might have increased by 4 to 5 times during the same period.

The gross income for an acre of cultivated land in Koraput district was Rs. 299 in 1977 and the figure stood at Rs. 1299 in 1987-88 as per the study of the current researcher. The value productivity per acre in Punjab. Tamil Nadu, Haryana and Andhra Pradesh was Rs. 3450.20, 3970.40, 2329.15 and 2414.57 respectively in 1988-89. The corresponding income for Orissa and All India was Rs. 1861.94 and Rs. 1859.88 respectively in 1988-89.

The Arithmetic's of the Table 12.1 shows that, both at country level and state level, there has been an upward revision in the procurement price of common paddy and superior paddy of FAQ from Rs. 150 per quintal and Rs. 158 per quintal in 1987-88 to Rs. 550 and Rs. 580 per quintal in 2003-04 respectively. The support price for wheat of (FAQ) has shot up from Rs. 173.0 per quintal in 1987-88 to Rs. 620.0 in 2002-03. And finally it stood at 630 per quintal for marketing season 2003-04. The procurement price has been fixed at Rs. 528 and Rs. 557-00 per quintal of common and grade of paddy in 2003-04.

Table 12.1: Procurement prices of paddy and wheat crops in Koraput district

Name of the Economy	*Reference year*	*Unit*	*Commodity*	*Price per quintal*	*All India average*
1	2	3	4	5	6
Koraput	1987-88	Quintal	Common paddy	Rs. 150-0	Rs. 150-0
Koraput	1988-89	Quintal	Do	Rs. 160-0	Rs. 160-0
Koraput	1987-88	Quintal	Fine variety paddy	Rs. 154-0	Rs. 154-0
Koraput	1988-89	Quintal	Fine variety paddy	Rs. 170-0	Rs. 170-0
Koraput	1987-88	Quintal	Superior variety paddy	Rs. 158-0	Rs. 158-0
Koraput	1988-89	Quintal	Superior variety paddy	Rs. 180-0	Rs. 180-0
Koraput	2003-04	Quintal	Common paddy	Rs. 550-0	Rs. 550-0
Koraput	2003-04	Quintal	Superior variety paddy of FAQ	Rs. 580-0	Rs. 580-0
Koraput	1987-88	Quintal	Wheat	Rs. 173-0	Rs. 173-0
Koraput	1988-89	Quintal	Wheat	Rs. 178-0	Rs. 178-0
Koraput	2002-03	Quintal	Wheat	Rs. 620-0	Rs. 620-0
Koraput	2003-04	Quintal	Wheat	Rs. 630-0	Rs. 630-0

Source: 1. Economic Survey 1988-89, Govt. of Orissa.
2. Economic Survey 2003-04, Govt. of Orissa.

Table 12.2: State-wise gross value productivity per acre in India 1988-89

State	*Value productivity per acre (Rs.)*
Tamilnadu	3970.40
Punjab	3450.20
Andhra Pradesh	2414.57
Kerala	2376.92
Harayana	2329.15
West Bengal	2248.58
Uttar Pradesh	2209.72
Gujarat	2037.65
Karnataka	2007.29
Jammu & Kashmir	1648.58
Orissa	1861.94
Assam	1483.40
Bihar	1455.46
Madhya Pradesh	1245.39
Maharashtra	1192.71
Himachal Pradesh	1207.29
Rajasthan	980.16
All India	1889.88

Source: J.L. Sharma.

Inter-state difference in value productivity, Yojana: Volume 35, No. 18, 15th/10/1991.

Legislation is a wonderful thing but in the absence of implementation, there is nothing to commanded it. The farmers of Koraput region are dragged to the whirlpool of distress sales of paddy due to reluctance of F.C.I. to buy paddy directly from growers.

As the editorial observation of the New Indian Express (23/12/2005) goes "despite minimum price fixed by government and a clear cut policy in force, it is the middle men who manipulate prices and make a huge profits at the cost of farmers who sell away his stocks at a throw away price. To begin with the government must set-up market yards in close proximity to ensure that farmers are not trapped by moneylenders and see that support price is paid by middle men and F.C.I."

SECTION-V

Policy Prescriptions

1. The study areas depict the use of arbitrary weights and measures like, stoned by traders for measuring the weightage of commodities. N. Senapathi and N.K. Sahu observe "Things are rarely sold by weight". So, with a view to ensuring consumer efficiency in provision of marketing services, the Government of India has promulgated Standard Weights Act in 1939 and old weight and measures was substituted by metric system.
2. Efforts may be made to ensure organised processing facilities in agriculture and related produces to enhance the share going to the peasants. To Koraput survey of 1987 by the scholar has surfaced that, farmers of Bhilaput and Kumbhari village of Narayanapatna block were compelled to sell tobacco to wandering traders at an unremunerative price of Rs. 970 per quintal far below the block level price of Rs. 1000 per quintal due to want to processing facilities for tobacco.
3. One sinister aspect of agricultural marketing in study areas relate to sale of adulterated food in "Hats". So, to give effect to grading of commodities and Enforce Consumer Protection Act, the quality control laboratory, numbering to 21 obtained in India are testing chemical properties of commodities like rice, wheat, cotton, ghee and vegetable oils. Grading of commodities is attended by the issue of "AGMARK" seal of authorised dealers.
4. There is an urgent need to establish regulated markets in the country. In line with the recommendations of All India Rural credit survey 1954, regulated markets were created both at national and state levels. Maintenance of data, settlement of disputes, quickening of payments to farmers produce, licensing middlemen's business and

lenders activities and specifying market charges are some of the functions entrusted to regulated markets. The performance of the aforesaid functions by regulated markets may accelerate the growth of the economy.

5. The production of farm products require credit, since farmer must incur expenses before he receives anything from sale of his output. The provision of cheap credit, will go a long way in containing the malignant influence of shy-lock like moneylenders and it may preclude the possibilities of distress sales by growers to moneylenders. Similar observations were made by world development report, 1986 "Farmers need better prices, easier access to inputs and lower market costs".

6. One palpable deficiency in the existing marketing system is the deficiency of road form of transport, official data convey that, there are 43 kms. of roads per 100 sq. kms. in the study area and these are not surfaced roads. It is regrettable that, all weather connection could not be provided to all villages having a population of 1500 and above, though the sixth plan had the vaulting ambition of connecting all villages, having a population of 1000 and above by All Weather Roads.

7. **Streamlining the working of LAMPS**

The scale of endeavours may be raised to fill in the gaps in the operation of LAMPS, such as, inordinate delays in delivery of inputs, lack of trained officials, want to ›rage facilities and official failure to make recurring › to inaccessible villages for purchase of agricultural ›ts from growers.

›opment related activity taken up by the centre ›sation of a Regional Agriculture Marketing Bhubaneswar. However, the existing ›sa are not considered for central › not possess a minimum of 15 acres ›ay be relaxed by the Centre, for a › Document, Government of

9. The other attributable factor for the lackluster performance of markets in study areas and Orissa is the inadequacy of storage facilities. The availability of storage facilities, doubtlessly may enhance the retention capacity of farmers and saves him from distress sales. The benefits of F.C.I. have accrued to developed blocks of Jeypore and have not tricked down to backward blocks of Narayanapatna. So, to facilitate the percolation of benefits of storage facilities over space, cooperatives and village panchayats may insure godown facilities to villagers.

10. There is much room for widening the frontiers of market research, extension and survey works in Orissa and India. Right from Fourth Five Year Plan, research endeavour in market sphere has commenced at National and State level. At present, there is only 1 marketing intelligence research officer assisted by two supervisor and one inspector, who has the onerous responsibility of disseminating data on supply, distribution and prices to rural folk and urban elite. Attempts may be made to create an additional post of market intelligence officer in the district.

11. **Creation of Growth Centres**

    The story of agriculture marketing in the resource rich backward district is woeful, due to the absence of proper market links and communication links between village panchayats and their respective villages within the panchayats and between panchayats and Block Officer (P.C. Mohapatra, Economic Development of Tribals in India). The villages are not linked to "Hats" so, each rural market, whether it is regular or weekly, covering a minimum of 15 kms. may be developed as a growth centre, if it has an attendance of 5000. Such growth centers should have infrastructural facilities like, All-weather roads, market yards, storage facilities, information cell and transport links. Market villages, with attendance of 3000-5000, traders, producers and buyers may be considered as sub-growth centres, with basic requirements, such as market yards, post office, cooperatives and information cell.

Conclusion

Marketing of agricultural products involves a constellation of forces like credit, storage, information, processing and transport facilities. All these activities allied to marketing may be considered in their totality, instead of considering marketing in a fragmentary manner. This calls for a high degree of coordination and concerted action on the part of extension agencies, input supplying agencies and public officials.

It is no gain say that, the need for comprehensive agricultural policy is pressing, when the country is looking for farming community to augment production to meet the needs of people. Efforts may be made to introduce comprehensive health insurance and social security measures, like Annapurna scheme, as Harayara has done. Marketing and pricing are weak links in our production programmes and much needs to be done to ensure fair price both to producers and growers.

REFERENCES

1. Singh R.I., Singh G.N. Singh R.K, and Singh R.L. "*Agriculture Problems and Marketing System*" Kurukshetra, June, 1985, Vol. 33 p. 47.
2. Cohen, R.L., "*The Economics of Agriculture*" James Nisbat & Co. Ltd., Cambridge, 1965, p. 77.
3. Uppal, J.S. "*Indian Economic Problems An Analytical Approach*" Tata McGraw Hill Publishing House, New Delhi, 1984, p. 77.
4. Lee Uma, "*Marketing Pricing of Foodgrains*" Paper in "Agricultural Prices and Economic Development by A. Srinivas Murty and Ramesh Verma, Himalaya Publishing House, Bombay, 1984, p. 137.
5. Raj Krishna, "*Agricultural Prices Policy and Economic Development*" by A. Srinivas Murty and Ramesh Verma, Himalaya Publishing House, Bombay, 1984, p. 62.
6. *Orissa Development Report 2002*, Planning Commission N.K.C. Centre for Development Studies, Bhubaneswar.
7. Tripathy S.N., "*Cooperatives for Tribal Development in Orissa*", Editor S.N. Tripathy" Tribals in Transition, Discovery Publishing House, New Delhi-1999, p. 88.
8. Government of Orissa, *Sixth Five Year Plan 1980-1985*, Bhubaneswar.
9. *World Development Report, World Development Bank,* Oxford University Press, London, 1986 (p. 7).
10. Sunder Rajulu—*NCDC Bulletin*, April 1985, Bombay, p. 88.

11. Dutt Rudar A.K.P.M. *Sunder An Indian Economics,* Sultan Chand & Co. 2005.

12. V. Padmanabhan, *Hindu,* May 20, 1990, p. 23.

13. Unpublished Thesis of Patnaik Eswar Rao *"The Problems and Prospects of Agricultural Development during the Sixth Five Year Plan in Koraput District",* 1992, p. 252-238.

14. Senapathi N. and Sahu, N.N. *Orissa District Gazetters"* Orissa Government Press 1966, Cuttack p. 216.

13

Watershed Development, People's Participation and Community Organisations

A Study in Kandhamal District (Orissa)

*Govinda Chandra Panda**

Introduction

India has traversed a long distance on the road to development since independence, achieving significant success in various sectors. But, the country has still to do a lot to make life better in tribal regions. The economic status of the tribal regions is in low profile. $2/3^{rd}$ of our tribal population are below the poverty line, $3/4^{th}$ does not have access to proper sanitation, $1/3^{rd}$ does not have provisions of safe and clean drinking water and over 1/2 is illiterate. The deplorable economic environment now has left hundreds and thousands of tribes impoverished, subjugated and vulnerable. Further, the economic reforms of early 90's have deprived the tribes either partly or completely to exercise their rights over the traditional natural resources.

Since independence crores of rupees have been invested and plenty of experiments have been made still the basic contradictions, which were affecting the causes of low sustainability of key economic institutions have continued to prevail even today. Availability of funds is not a major constraint in tribal regions. Obviously there are

* Mr. Panda is Head, Department of Economics, AMCS College, Tikabali, Kandhamal, Orissa.

some other barriers. Is it gap in mindset? It is wrong in commitment? Is it deficiency in prioritization? Is it unsuitability in technology adoption? Is it non-participation of the people in resource management? Certainly the answer is all. But the most worrying problem is the non-participation of people which cause serious concerns in the management of the natural resources in tribal regions. Water, a fundamental natural resource, is said to be the ecological base of all life. It is a common knowledge that water resources hold the key for socio-economic development of any country. It determines the quality of environment as well as the existence of our bio-system. But it is also a fact that water, a fragile and finite resource, is rapidly becoming scarce day-by-day. High consumerism, over exploitation, wrong management and inadequate policy and regulation now, have resulted in fast shrinking of these resource in tribal regions, thereby, causing serious threats to their survival.

Hence it is quite pertinent to address the issues of water crisis and its management with the participation of people in tribal Kandhamal—so as to make the approach holistic and participatory in policy formulation. In this backdrop attempts have been made in this research write up to explore the following objectives.

Objectives of the Study

1. To study the problems of water crisis in the agro-forest economy of Kandhamal District (Orissa);
2. To highlight the degree of participation of the people in watershed development (WSD) for sustainable water management;
3. To examine the institutional arrangements, funds allotted and expenditures incurred towards WSD in the study district;
4. To form a futuristic overview of plans and prospects of tribal development.

Methodology

Keeping in view the above objectives the district of Kandhamal in the State of Orissa has purposively been selected for the study because of its peculiar feature of heavy rains in rainy seasons and complete drought in summer days.

Group discussions and observation methods have been used to collect information from the field. Individual interactions and spot visits have also been made by the researcher to make the study more realistic. Both primary and secondary sources of data collection have also been taken resort to.

Water in the Agro-Forest Economy of Kandhamal

The geo-climatic situations of the district largely determine the availability of ground and surface water; hence, it is quite imperative to focus these situations to discuss the issues relating to water crisis in Kandhamal.

The Geo-Climatic Situations of Kandhamal

The total geographical area of the district is 7650 Sq. kms. The district is centrally located and lies in between 80° 30′ to 84° 35′ East Longitude and 19° 34′ to 20° 34′ North Latitude. The climate is sub-tropical and extreme in nature. Temperature varies from 0° Celsius to 45° Celsius. It is hot and dry in summer, humid monsoon, cold and dry in winter. Heavy cold spells in the entire district and show falling in Daringbadi and Kotagarh Blocks are common features of the district during the winter days. The district is situated at a height of 1700 ft. from the mean sea-level.

Topographically, the district exhibits hill-cum-forest tracts. High mountain slopes, plateaus, steeps, forest tracts, rocky surfaces, Dunes and Slacks are found in the district. The district manifests mostly sandy loam and red-laterite soils. Phyto-geographically, the vegetation of the district comes under two major forest divisions i.e. (i) Northern tropical semi-evergreen forest; and (ii) Northern tropical moist deciduous forest. The forests of Kandhamal is rich in flora (218 Nos.) and fauna (141 Nos.) The district also determines the Bio-sphere of the State in a larger extent. The demographical profile reveals that the district is mostly inhabited by Scheduled Tribes (58%) and Scheduled Castes (22%) population to total population of the district. The occupational structure reveals that the people of this district mostly depend upon agriculture and forest for their livelihood.

Water Crisis in Kandhamal: A Strategic Review

The district is in heavy rainfall zone because of huge forest coverage (4649 Sq. kms. including forest lands). The average rainfall of the district is 1576 mm. spread over 45 rainy days. It implies that rainfalls with high intensity but in short spell. Hence, water in this district needs to be preserve the otherwise the situation may lead to drought. But, the Water preservation and conservation capacity of the district is abysmally low. The fact is that the district is yet to have a major irrigation project. Even, the number of medium and minor irrigation projects can be counted in finger tips. The district agriculture report can well define the status of water provisions towards agriculture in the district. Irrigation facilities available for Khariff and Rabi are 10223 Ha. (8%) and 5111 Ha. (4%) respectively. In addition to this, agriculture is seriously affected because of low moisture and water carrying capacity of the soil as the soil is deficient in nitrogen, humus and lime. Again, the soil is light textured and acidic in nature having PH 5.3 to 6.3, which cause heavy water run-off and soil erosion in Khariff. The important point to note here is, out of 127790 Ha. available for cultivation in the district 96298 Ha. are High-lands. Ironically, these High-lands remain uncultivated through out the year. Even if cultivated rice cultivation is a nightmare. Thus, the vast quantity of water simply flows away to other neighbouring districts. Mountain slopes, steeps and plateaus also help water to flow with high speed. Again, the red laterate and sandy soils covered with rocky surfaces arrest water infiltration into the deep ground. It also prevents water re-charging. The result is acute shortage of drinking water in summer days. According to an estimation of D.R.D.A., Kandhamal.

(i) In an average water can be found for a well only after 45 feets. The cost of digging a well (5 ft. breadth) is around 40 thousands.

(ii) Lift irrigation points are hardly found. Tubewells catch water layer only after 250 mtrs.

(iii) The estimation further reveals that 42 per cent of tube wells are not able to produce water, 30 per cent tube wells produce water only for 1 to 2 years and the rest are producing for a maximum period of 7 to 10 years.

Thus, searching for drinking water in the district both at governmental and domestic front carry high risk of finance and uncertainty. Waterlogging due to forest tracts, dunes and slacks make it unsuitable for drinking. To add more to this problem, the inhabitants of the district hardly know the practice of water recycling. The water logged areas thus become a suitable breeding ground for malaria parasites mostly PFR (+). The district is one of the worst victims of water-borne diseases in the State having the infant mortality rate 116 per 1000. To conclude, the district is heading towards heavy water crisis. It is mainly for two reasons:

(i) Mis-management and lack of water conservation infrastructure;

(ii) Widespread deforestation of natural forest and its replacement with monoculture plantation.

Thus, the district of Kandhamal needs high attention in water management because the district receives huge waters in the beginning and becomes completely drought in the end.

Watershed Development in Kandhamal: A Case Study

Watershed development (WSD) was started in India in 1962-63. The Scheme "Soil conservation works in the catchments of River Valley Projects" was mainly done to control siltation of reservoirs. Basically, there were three approaches in WSD like Desert Development Approach (DDA), Command Area Development for Irrigated Area (CAD), Watershed Development for Rain-fed Area (WSD).

Watershed is a geo-hydrological unit or piece of land, which drains at a common point. In other words, watershed is a definite area which has an undulating terrain and from which the entire rain falling is draining out through a single outlet. This outlet may be a Nalla, a stream, tributary or a large river (Annexure-I). The WSD approach is aimed at augmentation and stabilisation of production and productivity. Hydrologically it is a correct approach for efficient use of soil and water for irrigation. Watershed is not a technology but a concept which integrates conservation, management and budgeting of rain water through simple but discreat hydrological units. Simultaneously, a watershed support to promote sustainable development, employment generation and ecological balance with the participation of people at the grassroots level. Watershed development, particularly in the district of Kandhamal is the only solution to ensure drought proofing and to mitigate the distress caused by frequent droughts. It is a right means for increasing and stabilizing production of crops, fruits, fuel and timber by improved soil and moisture conservation measures, scientific technology practices and reforestation. With this mindset attempts have been made to study the sustainability, rationality and viability of WSD in Kandhamal from two aspects:

(i) Institutional arrangements, people's participation and community organisations;

(ii) Release of funds and expenditures incurred.

1. Watershed Development, Institutional Arrangements and People's Participation in Kandhamal

The district-wise watershed, sub-watershed and mini watershed in Orissa (Annexure-II) clearly indicates that erstwhile Phulbani* has 39 watershed, 289 sub-watersheds and 889 mini watersheds covering a total area of 10,47392 Ha.

The coverage area is good enough to make strong provisions of water and irrigation facilities for agriculture in the district. As far as rationality and viability of WSD is concerned there are no serious threats in the study district. But, the sustainability of WSD largely depends upon people's participation and funds allocation.

(i) **People's Participation:** People's participation in natural resources management (NRM) is not concept in tribal societies. Historically, they are using forest and forest products for the benefits of the community as a whole. Centralisation and privatization are rather the headaches in common resource

---

* **Boudh, a Sub-Division, Earlier in Phulbani District, has been made a Separate District. Now the District is Called as Kandhamal.**

utilization in tribal regions. Thus the sustainability of water resource management can better be realized with people's participation in the study district of Kandhamal. The institutional arrangement of WSD in Kandhamal depicted in Annexure-III highlights that Collector/Chief Executive Officer as a mission leader is at the apex. The District Rural Development Agency as a chief implementing agency convenes the meetings of the District Water Advisory Committee (DWAC) and executes the proceedings at the district level. WDT and PIA's are the nodal agencies functioning at the Block level to give technical guidance's and training facilities to WA, WC, SHG, UG, WS & WV members. The institutions like WA, WC, SHG, UG WS & WV operating at the Grama Sabha and watershed level are the real stakeholders and decision making bodies (see Annexure-IV). The number of institutions, PRI's and NGO's* as PIAs* and persons connected with WSD has been illustrated in Annexure-IV (a & b). In addition to this, in accordance with 73rd & 74th Amendments to the Constitution of India and PESA* it has been investigated that Zilla Parisad, Panchayat Samitis and Gama Panchayats through Gram Sabha are actively participating in policy formulation execution, supervision and monitoring of WSD activities in Scheduled and Tribal reasons.

Each institution associated in watershed development in the study district has specific organisational arrangement and operational framework towards ensuring people's participation in the management and implementation of watershed programmes (WSP) at various levels.

The organisational mechanism of the institutions like WA, WC, SHG & UG manifest that people with different identities and exposures involve themselves in various groups:

*** PIAS- Programme Implementing Agencies, NGOs- Non-Government Organisations, PESA- Panchayat Extension Scheduled Area Act.**

**** WDT- Watershed Development Team, WA- Watershed Association, WC- Watershed Committees, SHG- Self Help Group, UG- User's Group, WS- Watershed Secretary, WV- Watershed Volunteers, WSP- Watershed Programme, UC- Urilizatioon Certificate.**

(a) towards the formation of community organisations; and

(b) towards the participation in community organisations.

Towards the Formation of Community Organisations

The participation of the people in WSP through community organisations has been depicted in the box-diagram illustrated below. The interlinking of people through community organisation with WSP begins from watershed levels to district level.

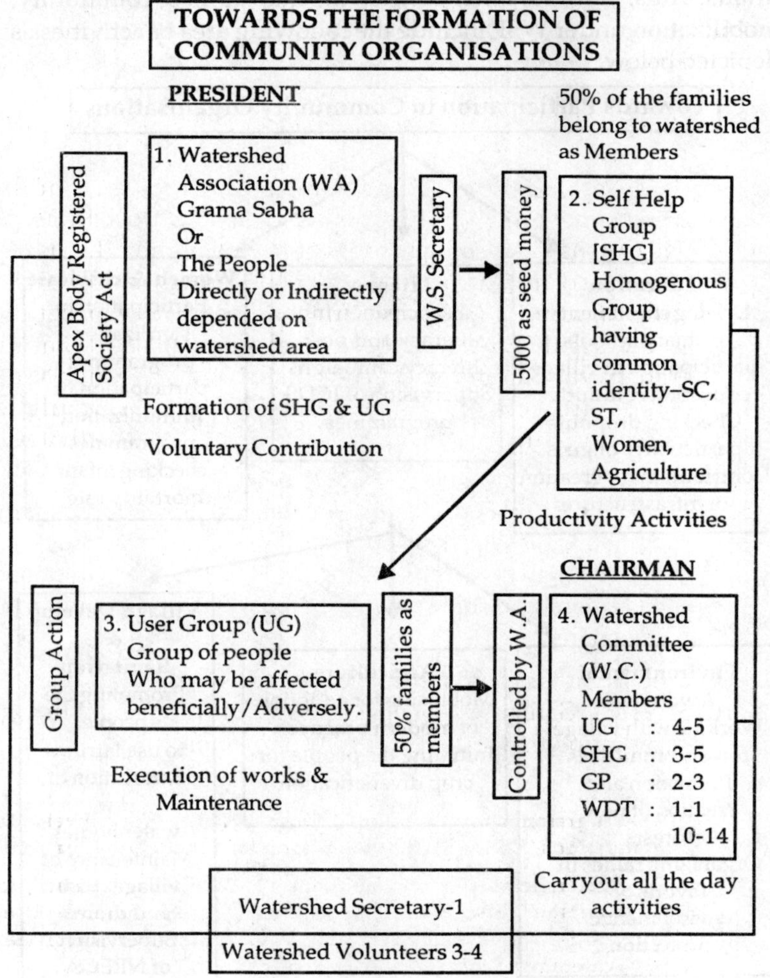

Towards Participation in the Community Organisations

The watershed association members are mostly participating in the community activities like—Organising SHG's/UG's, conducting participatory rural appraisal (PRA) exercises, conducting awareness camps cum literacy, family welfare, social service and income generating activities, extending financial helps to SHG's and UG's, repairing community halls, village roads and drains. Thus, in Kandhamal District, the indicators of community, mobilization under WSP include the following area of activities as depicted below:

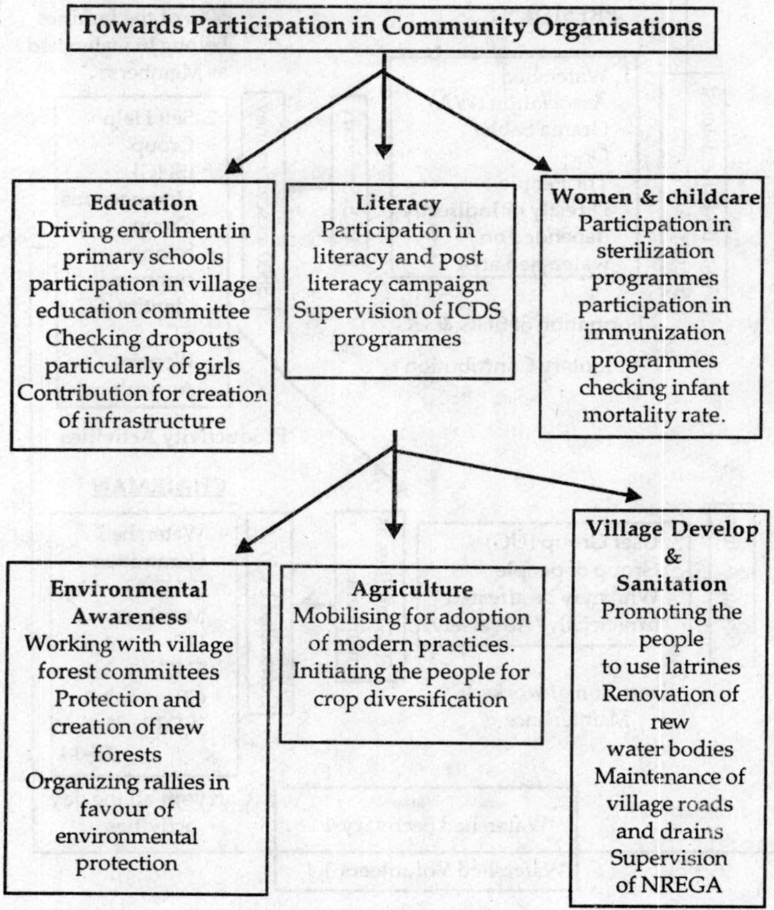

Expenditure and UC Position upto 31.10.2004—**DPAP 1st Batch**

Sl. No.	Name of the PIA	Total funds released to PIA & WA		Expenditure made	UC submitted	Balance	
						Fund	UC
1	2	3		4	5	6	7
		PIA	2179407	2179407	2179407		
1.	ASCO, Phulbani	WA	7171978	7171978	7171978		
		Total	9351385	9351385	9247922		103463
		PIA	819561	755467	181627	64094	573840
2.	BDO, Khajuripada	WA	5332220	5015598	2217509	316622	2798089
		Total	6151781	5771065	2399136	380716	3371929
		PIA	570511	564181	564056	6330	125
3.	AGRAGAMI, Phiringia	WA	2649262	2262297	2262297	386965	
		Total	3219773	2826478	2826353	393295	125
		PIA	1270288	1281188	1281188	-10900	
4.	NIPIDIT, Tikabali	WA	6615314	6577794	6554794	37520	23000
		Total	7885602	7858982	7835982	26620	23000
		PIA	1632162	1626787	1494875	5375	131912
5.	ASCO, Kotagada	WA	7275968	7275968	6471255		804713
		Total	8908130	8902755	7966130	5375	9366225

(Contd...)

Sl. No.	Name of the PIA	Total funds released to PIA & WA		Expenditure made	UC submitted	Balance	
						Fund	UC
1	2	3		4	5	6	7
6.	J.E., Tumudibandha	PIA	1741077	1713174	1713174	27903	
		WA	5793208	524875	524875	548333	
		Total	7534285	6958049	6958049	576236	
7.	Horticulturist, Baliguda	PIA	971550	869283	851080	102267	18203
		WA	5273751	5073251	5067651	200500	5600
		Total	6245301	5942534	5918731	302767	23803
8.	SWADESHI, K. Naugam	PIA	2711394	2711394	2711394		
		WA	8304455	8304455	8138528		165927
		Total	11015849	11015849	10849922		165927
9.	JAGRUTI, Daringibadi	PIA	1616727	1830907	1830907	-214180	
		WA	6772246	6412932	6412932	359314	
		Total	8388973	58243839	58243839	145134	
10.	ASCO, G. Udayagiri	PIA	175471979	1754719	1754719		
		WA	45452	7945452	7945452		
		Total	9700171	9700171	9700171		

(Contd...)

Sl. No.	Name of the PIA	Total funds released to PIA & WA		Expenditure made	UC submitted	Balance	
						Fund	UC
1	2	3		4	5	6	7
11.	PRADATA, Chakapada	PIA	2008759	2100964	1757788	-92205	343176
		WA	6412119	6371777	6361776	40342	10001
		Total	8420878	8472741	8119564	-51863	353177
12.	BDO, Raikia	PIA	50000	50000	31065		18935
		WA	2157217	2150258	2040113	6959	110145
		Total	2207217	2200258	2071178	6959	129080
	Grand Total		89029345	87244106	82136977	1785239	5107129

Source: Proceedings of the review meeting on watershed development.

Expenditure and UC Position upto 31.10.2004—**DPAP 9th Batch**

Sl. No.	Name of the PIA	Total funds released to PIA & WA		Expenditure Made	UC submitted	Balance	
						Fund	UC
1	2	3		4	5	6	7
1.	N. Patro, Phulbani	PIA	300000	113981		186019	113981
2.	S. Dixit, JSCO, Khajuripada	PIA	300000	149000		151000	149000
3.	K.P. Rot, JE, Phiringia	PIA	600000	369500	200000	230500	169500
4.	A. Mishra, JSCO, Chakapad	PIA	300000	130000	80000	170000	50000
5.	C.S. Sahu, JSCO, Kotagarh	PIA	300000	20000		280000	20000
6.	P.K. Samantaray, JE, Tumudibandha	PIA	300000	19000		281000	19000
7.	J.N. Mishra, PIA, Balliguda	PIA	300000	35000		265000	35000
8.	A.Rath, JSCO, K. Nuagam	PIA	300000	80000		220000	80000
9.	S.R. Rout, Daringibadi	PIA	300000	33278		266722	33278
10.	P.C. Dash, Raikia	PIA	300000	166200	120000	133800	46200
11.	H.H. Behera, JE, Tikabali	PIA	300000	45000		255000	45000
	Grand Total		3600000	1160959	400000	2439041	760959

Source: Proceedings of the review meeting on WSD.

(ii) Allocation of Funds and Expenditures Made

Annex-5 (a) demonstrates the pattern of funds released to PIA's WA and PRI's in the study district. Funds are mainly released from the DRDA to the respective PIA's. It has been found that the funds have been released to PIA's like Panchayat Samities. NGO's Government Line Departments, Watershed committees and Gram Panchayat in the district. The expenditure of funds include the improvement and establishment cost of WSD, salary of the WV and WC, office maintenance cost and training expenses of WA, WC, SHG's and UG's members. Total funds released, expenditures made and UC submitted have been manifested in the following tables of DPAP 1st Batch and 9th Batch.

Remarks and Policy Options

The remarks and policy options have been suggested here on the basis of observation and investigations. As a part of O & I the researcher used to visit. The spots of WSD projects and extensively utilized the records of the office of DRDA, Forest Division and Soil Conservation of Kandhamal District to study the realities in the ground level. WSD programme has both progression and retrogression. These are summed up as follows.

Positives

1. It has created sustainable livelihood and employment opportunities in the form of man-days for the poor tribes.
2. It has checked the problem of labour migration as depicted in District Statistical Hand Book Kandhamal.
3. The tribes have been made aware to exercise their rights over the community resources as manifested in Annex-IV (A & B) in the form of WU and WD committees.
4. The adjacent highlands have been brought into cultivation. District Agricultural Report reveals that area of irrigation has increased to 5 thousand to 5 thousand Ha. for Rabi and 8 thousand Ha. for Khariff.
5. The user groups and water committees of WSD are actively engaged in forest protections where village forest committees are not functioning as is observed.

6. Women SHGs are participating actively in the campaign of use of safe drinking water.
7. Soil erosion has been checked considerably. Moisture retenvity and water carrying capacity of the soil in WSD areas have increased consistently. (Report of Soil Conservation).
8. As per the official records of DRDA, due to optimum interaction of water and soil, the recharging and harnessing of surface and sub-soil have been made possible.
9. It has considerably checked pollution caused due to depletion of forest resources.
10. It has been observed that there is substantial improvement in the ground water table.

Negatives

1. The PIAs like ASCO Phulbani, NIPIDT Phulbani, ASCO G.Udayagiri, PRADATA G. Udayagiri, BDO Raikia, Jagruti Daringbadi have not submitted the list of projects executed and expenditure made in given proforma. (Proceedings of the Review Meetings of WSD).
2. PIAs also fail to submit the documentations and survey reports with reference to Bench Mark data. (Proceeding of the review meeting on WSD).
3. It has been found that inspite of DRDA Letter No. 3941 dated 11.08.2004 Mid-term evaluation reports have not been submitted by the PIAs like ASCO Phulbani, NIPIDT Phulbani, ASCO G. Udayagiri and SWADESHI Phulbani.
4. PIAs are not submitting the success stories with photographs.
5. PIAs are not submitting utilization certificates properly.
6. Huge amounts of funds remain balance.
7. Monthly progress reports of SHGs UGs are not furnished.
8. No PIAs have submitted the reports about the construction of green poly house.

9. BLWIRC meeting at the Block level are not convened to discuss progress of watershed works by the PIAs.
10. The PIAs like JE phiringia, JE Raikia have not submitted the information on BPL households under SHG in Watershed area.
11. Corruption at all levels seems to be a big challenge for success of WSD Programme in Kandhamal. (Paper cutting pasted).
12. People's voice are not reflected in some instances. (Paper cutting pasted).

Conclusions and Suggestions

The ever-increasing mis-match between supply and demand scenarios has aggravated the situation in the form of falling water levels, degradation of water quality and concomitant breakout of water borne diseases. Thus, it needs a solution both from demand-supply aspects. The demand-supply solution to water is not obviously from market point of view. Market solution may destroy the life and ecological base of the earth. The causes of Water crisis may be commercial, but when water disappears, there is no alternative. It is a free gift of nature; available partly in store and partly in flow. To Vandana, S. Water Scarcity means for women of Third World, large distance in search of water. For cultivators it means starvation, destitution and suicide as drought wipes out their crops. For children it means dehydration and death. Hence market and market forces are not the solutions in the adjustment of demand and supply of water. It is not tradable. Application of Market Economics is not an answer to it. It rather needs socio-economic and legislative safeguards. The safeguards include conservation and storing, re-charging and recycling, harvesting and treatment, efficiency in water use and development of cultures of water use and finally harnessing of unutilized water resources. Legislatively, there should be provisions of strong punishments for wrong users of Water. Appropriate judicial steps should be taken against the industrial houses causing serious water pollutions.

As far as Kandhamal district is concerned the serious bottlenecks of development are unconquered poverty, unwieldy unawareness, inaccessible illiteracy, uncontrollable population growth, unparallel disequilibria and irremovable inequalities. These problems must be tackled to bring them into the developmental mainstream. Secondly, in the process of natural resource management the opinion of the tribes must be sought, so that, "Participatory Management" can be made more meaningful. Thirdly, the planning must start from the people in order to make it a people's movement. The activities of the Governmental bodies should be restricted in the levels of supervision and guidance. Fourthly both feasibility and sustainability of WSD need to maintain by the way of post project evaluation to meet the varied economic needs of the tribes of this district Fifthly, experts from agriculture and plant sciences should be engaged to monitor the progress of the projects. Sixthly eco clubs, student organisations, youth clubs and teachers should be included during the training periods of watershed members. Seventhly, corrupt officers should be punished without delay, so that transparency and credibility of WSD can be realized. Seventhly, serious attempts should be made to preserve and conserve water by applying both indigenous technology (IT) and modern scientific knowledge (MSK). Eighthly, deforestation and replacement of natural forces with monoculture plantation must be avoided to check uneven and inconsistent rain falling.

Playing with water means serious threats to the survival of living biotic organisms. To conclude, the prediction of Sunderlal Bahuguna "the acute scarcity of water may force the powerful nations to wage a new global war for the control of the depleted sources of water" may also come true. So we should respect the United Nation's Message "Water is a scarce resource to be managed and protected."

WATERSHED

ANNEXURE-I

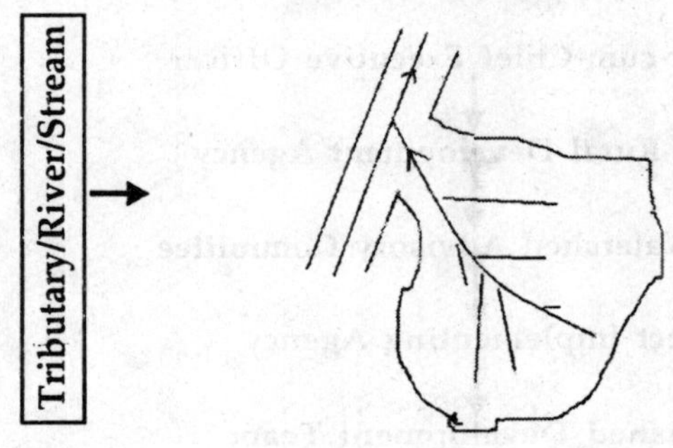

ANNEXURE–II

Sl. No.	*Districts Undivided*	*Watershed*	*Sub Watershed*	*Mini Watershed*	*Total Area in Ha.*
1.	Mayurbhanja	55	292	785	1042134
2.	Balesore	60	190	493	426080
3.	Keonjhar	61	173	409	830300
4.	Cuttack	61	216	530	1114200
5.	Dhenkanal	59	310	965	1082700
6.	Sambalpur	102	535	1423	1751600
7.	Sundargarh	100	341	846	971200
8.	Koraput	78	676	1531	2105253
9.	Kalahandi	42	318	781	1183589
10.	Bolangir	19	91	289	883000
11.	Ganjam	86	276	684	1234945
12.	Puri	9	27	88	719204
13.	**Phulbani**	**39**	**284**	**889**	**1047392**
Total		**771**	**3729**	**9663**	**14391597**

Source: User's manual for watershed development, Govt. of Orissa, January, 2001.

ANNEXURE-III

(Institutional Arrangements in WSD Programme)

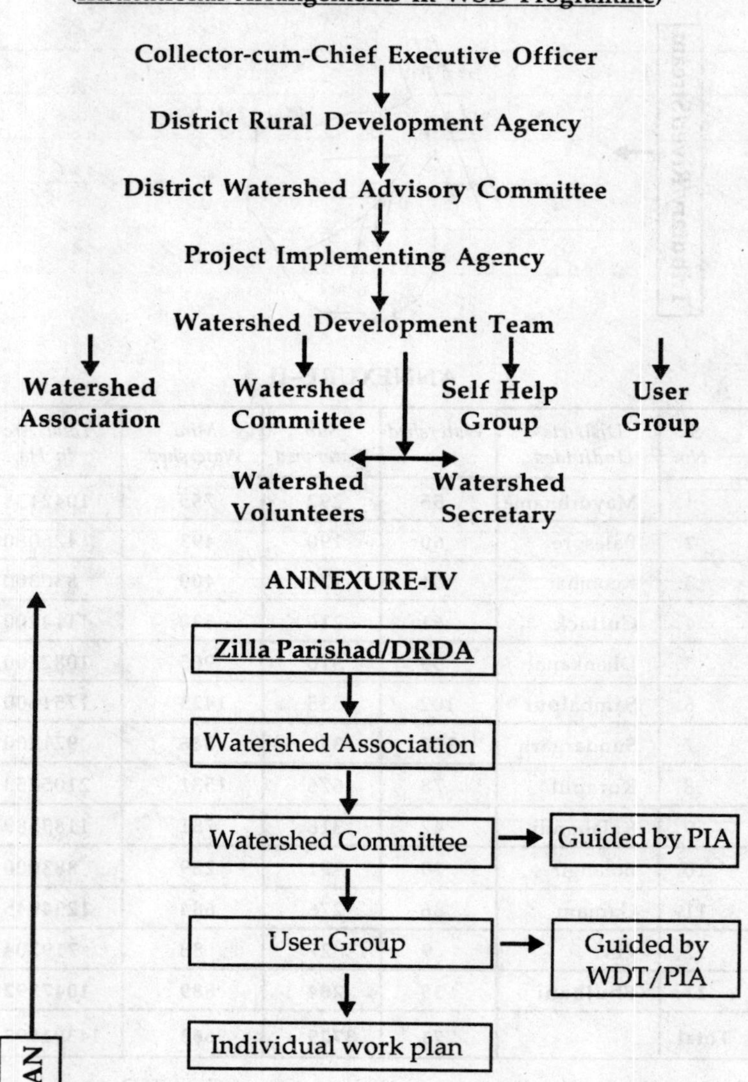

ANNEXURE-IV (A & B)

(People's Participation in WSD Programme)

4 (a)	Sl. No.	Name of the P.I.As
	1.	SWADESHI
	2.	PRADATA
	3.	NIPIDIT
	4.	JAGRUTI
	5.	AGRAGAMI
	6.	ADAO
	7.	PANCHAYAT SAMITIS
	8.	ASCO

4 (b)	Sl. No.	People's Organisation	In Nos.
	1.	WA	700
	2.	WC	1475
	3.	SHGs	4674
	4.	UG	1872
	5.	WS	1176
	6.	WV	14176

Source: Office of the DRDA, Phulbani.

ANNEXURE-V

Release of Funds

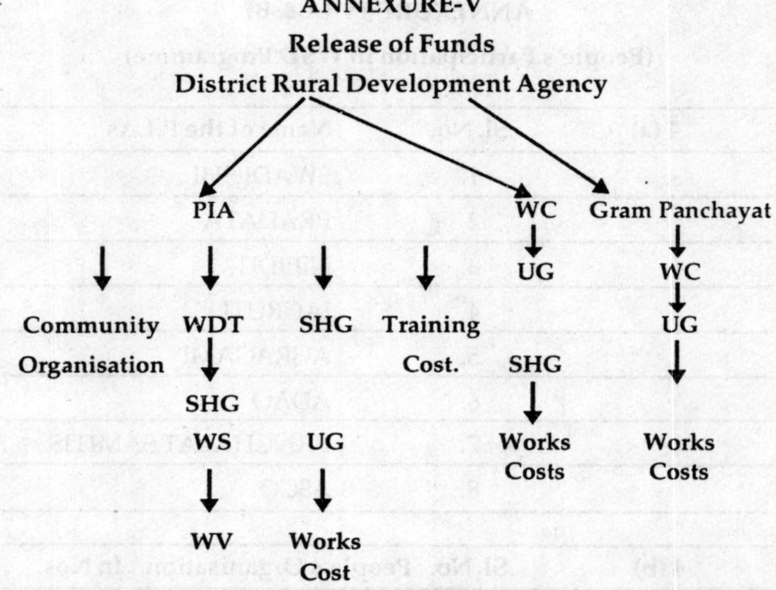

Source: User's manual for watershed development, Govt. of Orissa, January 2001.

REFERENCES

1. Gosh, S. and Verma, H.N.—"*Water in 21st Century.*" Kurukshetra Vol. 50 No. 3, January 2002.
2. Meena O.P., Charak A.S. & Gour, B.L.— "*Global Warming and Its Impact on Agriculture.*" Kurukshetra Vol. 53, No. 8, January 2005.
3. Bandana, S.— "*The First and Last Commons*"— Yojana, Vol. 49, June, 2005.
4. Kalam Abdul A.P.J. — "*Integrated Water Mission*" Yojana, Vol. 49, June 2005.
5. Hedge, P.— "*Linking Forests and Rivers*" Yojana, Vol. 49, June 2005.
6. Duggal V.K.— "*Drinking Water and Sanitation*" Yojana, Vol. 49, June 2005.
7. Panda, G.C. "*Development of Tribal Women Through Self Help Groups: A Study in Kandhamal District*", Paper Presented in an UGC National Conference (8-10 August 2004), Nagarjuna University, Guntur, AP.
8. Panda, G.C. "*Empowering Tribal Women Through SHGs: A Study in Kandhamal District*" Paper Presented in An UGC State Level Conference (12-13 September 2004), UNS College, Khairabad, Mugpal, Jaipur, Orissa.

9. Panda, G.C. & Tripathy, Dr. S. "*Impact of Rural Transport on Population: A Study in Kandhamal District, Orissa.*" in "Problems of Population in India" Edited by Panigrahy, R.L., Discovery Publishing House, New Delhi, 2004.

10. *Smaranika Kandhamala Mohouchhaba 2003*: District Council of Culture, Kandhamal District, Phulbani.

11. *District Statistical Handbook Kandhamal-1999*: Directorate of Economics and Statistics, Govt. of Orissa, Bhubaneswar.

12. *Official Records of DRDA*, Soil Conservation, Agriculture and Forest Departments.

14

Ensuring Sustainable Rural Development Through Participatory Management of Water Resources

*Krishna Chandra Pradhan**

Water scarcity has emerged especially during the past decades as an important theme of global discussion. It is realized that, water is going to be a precious commodity which could lead to conflicts between nations. On world water day the United Nations has launched the international decade for action "Water for lLife" (2005-2015). The goal of the first decade (1981-1991) was to assure access to potable water to all inhabitants of the planet by 2000. As we know, this goal was not achieved. This failure was due to the fact that the leader of the rich and powerful countries did not implement the measures that they themselves considered essential, especially allocating 0.7 per cent of GDP per annum for aid to UDCS.

Thus in the year 2000, there were 1.5 billion people without access to clean drinking water and 2.4 billion without sanitation; between 15000 and 30000 people died each day from illness as a result—the equivalent in victims between 26 and 52 tsunamis each year. Faced with a human disaster of scale, the words heads of the states met in New York in September 2000 for the United Nations Millennium Summit, where they approved the Millennium Development Goals (MDG). Though in spirit this project was

* **Lecturer in Economics, Khemundi College, Digapahandi, Ganjam (Orissa).**

particularly ambitious. Infact the leaders abandoned the objectives of eradicating poverty and consequently the lack of access to clean water. Instead, they asserted that the only realistic major goal was to cut by half the number of people in extreme poverty and lacking access to water sanitation by the year 2015. infact the world leaders accepted that by the year 2015 there will be more than 3 billion poor and between 1.5 and 2 billion people without potable water—this in a world in which total wealth more than doubled between 1975 and 2000 and will be 1.5 times higher in 2015 than in 2000.

Water Poverty:— "Water poverty", a phrase used to indicate difficulty people face in securing adequate and reliable access to water for productive and consumptive uses. A related concern is also deterioration of water environment, reflected in drying up of wet lands deterioration of water quality and desertification. Global discussions over the past decade have resulted in a variety of views points about how best developing countries can cope with this imminent condition. At one extreme, researches suggest that crying need in for honing even more than in the past the social capacity of communities and societies to adopt to water security [Ohlsson and Turton 1999; Wolfe and Brookes 2003.] However a popular view is the urgent need to make a transition from water resource development Mode to water resource management mode by embracing Integrated Water Resource Management (IWRM).

The water poverty index (WPI) covering 147 countries published by researches from Keele University and Centre for Ecology and Hydrology, Wallingford, UK in 2003 provides a ready-made global data base to explore the relationship between water poverty and water scarcity [Lawrence, Meigh and Sullivan, 2003]. It is suggested that the real indicator of water poverty is "Water Access Poverty (WAP)". Water access poverty index in strongly related to Human Development Index (HDI). The higher the HDI of a country, lower the water poverty regardless of a country's water endowments. Thus water poverty is the result of water scarcity and the failure of water institutions and policies to counter it, which can be combated through IWRM.

The Global Water Partnership (2000), defines IWRM (Integrated Water Resource Management) as a "process which promotes co-ordinated development and management of water land

and related resources in order to maximize the resultant economic and social welfare in an equitable manner without compromising the sustainability of vital Eco Systems "The IWRM package" has basically included a clutch of following instruments:

- A national water policy so that in a cohesive, well understand normative frame work to guide all players in the sector;
- A water law and regulatory frame work for coordinated action for sustainable water resources management;
- Recognition of the river basin as the unit of water and land resources planning and management;
- Treating water as an economic good by pricing water resources as well as services especially outside life line uses, to reflect its scarcity value so that it efficiently used and allocated to high value uses.
- Creation of water rights, preferably tradable by instituting a system of water withdrawal permits;
- Participatory water resources management with involvement of grassroot institutions and women so that water becomes everybody's business.

People's Participation and the Role of Grassroot Institutions

Participation has acquired a wider dimension with the assertion for democracy participation is also associated with political liberties and civic consciousness. Participation breeds the ground for empowerment, good governance and capacity building. Simultaneously community assets building can have a positive impact on rural development.

Grassroot institutions like Panchayat Raj Institutions (PRIS), Non Government Organisations (NGOs), Self Help Groups (SHGs) and other social institutions have become hall mark of rural development in the present globalised world shaping the local institutions for the beneficiary goals in the element of self expression amounting to participation. Here participation is generated through a dialogue among the stack holders seeking the shape of the local facilities, similarly these institutions are the major agents a social change and can play an important role in the management of natural resources at the local level.

The participatory role of all people from top to bottom is crucial to successful initiation/design/financing/implementation of development programmes. The need of the hour is to prepare and activate local rural inhabitants capable of handling the environmental problems as well as the problems of over utilisation of natural resources.

People's Participation and Development of Watersheds

Participatory approaches are being tried in almost all sectors. The application of participation in rooted in the logic that people should have an opportunity for controlling their own fate, as much as possible. Participation has become virtually synonymous with development projects. It also serves as a vehicle for better delivery of services. A popular vehicle for people's participation and for delivering services in natural resources management is the users groups. Thus water user's association forest users groups and watershed committees among others have been formed all over Asia either as a complement to or a substitute for, centralised government controlled management.

The natural resources can be managed better by managing the watershed areas as a whole. Watershed, a natural entity itself, combines forest management, land use management and water management. It happens to be such a hydrologic or physiographic unit that if offers a very good unit for planning and implementing soil conservation, water harvesting, afforestation and environmental protection programmes. Watershed programmes can be effectively implemented with the active involvement and co-operation of the beneficiaries.

It is the call of the time that people should realise the intrinsic relationship between three 'Ps' (population, poverty and pollution). The poor, in our country are left with no option but to degrade their own environment for their own survival. Still, it is only they who can restore the health to environment thus; ruined; outside actors can only facilitate but never substitute for stake holders.

Development of Local Institutions as Manager's of Watersheds

Throughout human history water resources have been managed as common properly, but now privatization of water resources in gaining momentum, prior to the arrival of the British in

South India communities managed water systems collectively though a system called Kudimaramath (self repair). Under this system a peasant was required to pay 30 per cent of the grain which he or she earned to a public fund for the maintenance of water ponds and public works. The system was based on persuasion rather than coercion.

N.S. Jodha's study reveals that most of the basic needs like food, fodder, fuel, water etc. needs of the poor in India are satisfied by self managed cell regulating village community councils. In Rajasthan village community councils appoint their own watchman so that no community member or outsider breaks the self imposed rule.

The traditional water systems based on local management were also the insurance against water security in drought prone areas of Gujarat. These systems were managed by local institutions mainly by the village communities. In the event of floods famines and other calamities the king also helped.

Local institution in water management included farmers associations, local irrigation functionaries and the village water associations maintained by contributions from each family.

In Karnataka and Maharashtra the farmers associations for construction and maintenance of water systems were called as Panchayats. In Tamilnadu, they were called Nattamai, Kavari, Maniyam, Nir Maniyam Oppidi, Sangam or, Eri Variyam (Tank committee). In the Himalayas the irrigation managers were called Kohlis. Collective labour investment was the primary investment in the construction and maintenance of village water systems. Each able-bodied person was required to help maintain and clean channels. Kautulya's Anthashastra included certain punishments for defaulters from any kind of co-operative construction.

The 73rd and 74th constitutional amendments of modern India gave a new approach to development and social change. The amendments aim at democratic decentralisation and self governance. Thus PRI's have been the prime interments of decentralisation at the grassroots level in India.

As per the guidelines for watersheds developments (revised–2001), the Zilla Parishads and other Panchayat Raj Institutions

(PRI's) shall have very important role to play in watersheds development programmes. At the village level, the Gram Panchayats shall be fully involved in the implementation of the programmes.

It can be recognized that there are at least ten levels of decision making bodies that range from the international level down to the level of individual decision makers, viz.:

(i) International level
(ii) National level
(iii) Regional level
(iv) District level
(v) Sub-district level
(vi) Panchayat level
(vii) Community level
(viii) Group level
(ix) Household level
(x) Individual level

The former five's are no way local in the true sociological sense and the latter five requires power and capacity building for watershed projects, capacity building and community participation requires Group formation/Institution building; Skills for fostering group participation especially women groups; conducting participatory planning session, approaching local community etc.

Self Help Group (SHGs)

The SHG is considered as a viable organisation of the rural poor especially women. Women are the agents of natural resources management as they are in regular interaction with land, water, fuel fodder, food, livestock, medicinal plants etc. Therefore the SHG's can play a crucial an important role in watershed management. Each project implementation agency (PIA) is required to consult watershed development team (WDT). WDT is a multi disciplinary team handling 10-12 watershed each. Each WDT should have one women member at least. At the same time. The PIA shall also constitute User Groups (UGs) in the watershed area with the help of WDT.

Problems Associated with Planning and Implementation of Watershed Programmes

The watershed programmes are implemented by the Zilla Panchayat through watershed associations. A Project Implementing of Agency (PIA). Which may be a Government Department on an NGO, in assigned about 10 micro watersheds, each micro watershed covering about 500 heaters. The PIA forms a watershed development team that interacts with the watershed associations and provides technical assistance to the watershed association in the planning and implementation of the watershed programmes. The residents of the area covered by the watershed are also organised in the self-help groups and user groups. In fact these user groups are the beginning point as well as the end point of watershed development programmes. Their initiative is crucial to the success of the programme and they are ultimate beneficiaries.

It is observed that watershed projects can not succeed with out full participation of project beneficiaries and careful attention to issues of social organisations. This is because their success depends on consensus among a large number of users. Moreover, collective capability and action is required for management existing as well as new structure created during the project. Otherwise the coast and benefits of watershed interventions are location-specific and unevenly distributed among the people affected.

The unfortunate truth today is that most projects have failed to generate sustainability because of the failure of government agencies to involve the people.

This lack of sustainability can be attributed to a number of practices followed by the government in the implementation of watershed development programmes. Strict orientation to achieve physical and financial targets discourages the project authorities to promote peoples participation. There is considerable pressure to spend substantial recourses by a fixed deadline leaving no room for pursuing participatory approach. At times people participate by being told what has already been decided. Announcement on projects is made unilaterally by the government agencies without listening to people response. The information collected belongs to external professional; it is not shared with the people. It is imperative for the successful implementation of the watershed projects that people participate in the planning and execution of the project right from the beginning. This would inculcate a sense of ownership among the people. Some other bottlenecks of these programmes are:

- In the present form, schemes are planned and executed by district level officers who have very limited capacity to do so;
- No arrangement for handling over of structures and maintenance of plantation after a project in completed;
- Little evaluation of the programmes is done after it has run for a couple of years as it is taken for granted that once money has been spent, physical progress automatically results. But, this is far from the truth;
- Forest lands, non-forest pastures, waste lands and crop lands must be looked at in an integrated manner. This is usually not done;
- Programmes are run departmentally with little vertical compartmentalization. Horizontal linkages between various agencies are very week. Thus although watersheds development may require integration of soil conservation techniques with plantation, there is little likelihood of effective co-ordination between the District Soil Conservator and the District Forest Officers.

Strategies to Develop Successful Partnership

Participation, partnership and people (three P's) are the major concern for the management of watershed. It takes time and skill to create successful partnerships. Maintaining motivation and enthusiasm is another challenge. The stakeholders are the right partners for success of management of watersheds. Success depends involving a good mix of people and organisations in the partnership to put together and implement the plan. People are required to play a number of roles; viz. Technical leadership, communications, political liaison public policy etc. some people who live outside the watershed may even have an important role to play because they benefit from or impact water or other natural resources within the watershed. Here are some specific strategies to develop successful partnership.

- *Local Government Agencies:* Financial and technical support. Policies and decisions that affect the watershed. Logistics, equipment's and related support, data collection and analysis expertise.
- *Mass Media:* Coverage of watershed events human interest stories understanding of local information needs, ability to get information out quickly.
- *Financial Institutions:* Influence over management decisions, linkage with land owners, prestige for partnership, funding for programmes.
- *Local Elected Officials:* Political leadership and credibility. Land use and resource management decisions. Financial support for projects.
- *Environmental and Conservation Groups:* Knowledge of environmental constituencies, awareness of problems and issues. Committed and knowledgeable memberships.
- *Teachers:* Influence over values and beliefs. Ability to shape future generation's source of information.
- *Students:* Influence over efforts in the future. Time and energy for repetitive tasks.
- *Women's Groups:* Influence over family decisions interest and concern for health issues. Ability to mobilize and motivate members.
- *Religious Leaders:* Commitment to stewardship ability to appeal to higher values, credibility and legitimacy.

Path of Partnership

The success of partnership involves certain stages. This stage can be compared to how we learn to swim. Each stage is related with certain feeling and actions. The success is closely associated with understanding and preparing for these stages.

(a) *The Forming Stage:* People are hesitant swimmers and they stand by the side of pool and stick their toes in the water. Excitement and the optimism mixed with skepticism and anxiety one the basic feelings of the people in this stage. The chief activity into the define the job and discuss how to accomplish it.

(b) *The Storming Stage:* This is the most crucial stage. Partners start loosing patience and begin to argue. They are like new swimmers. After they jump in the water, they are afraid they might drown. They start arguing about less important issues and unrealistic goals.

(c) *The Normalising Stage:* Conflicts are reduced and competitions become more co-operative like experienced swimmers, people realise that they are not going to drown and they help keep each other afloat. Achieving harmony by avoiding conflicts and sharing of problems in the chief element of the stage.

(d) *The Performing Stage:* People begin to really work together, like a winning relay team, the partnership works together well. The chief activities of this stage are, constructive change, ability to work through problems and closer attachment to the partnership.

Measures to Strengthen Partnership

The following remedial measures are suggested to overcome the difficulties of partnership and develop healthy environment for participatory management.

- Select partners based on existing and potential skills, not personalities partnerships will need technical, communication, problem solving and interpersonal skills.
- People are often skeptical at the initial stage, so paying regular attention to early meeting.
- A bottom-top approach will work to a great extent. Leaders should emerge from among the members of the partnership. Some one will have to take the initial responsibility for getting members together. So there is the need to train both government officials and functionaries of agencies.
- Initialing the project with a few short-term goals has a good chance of success. Early projects are realistic and will be seen as winners in the eyes of the partners.
- There are three distinct groups in a watershed: (a) Those who are both affected by and interested in watershed;

(b) Those are not affected but interested; (c) Those who are affected but are not interested. All partners should be able to take part in any decision or activity where they have interest or expertise.

- Make sure that all partners are comfortable in placing their statements. Potential conflicts need to be discussed and resolved. It is important to keep the statement general enough to encourage wide spread support, but specific enough to identify goals and measure progress.
- Effective decision should not be taken to make some one happy. Most effective decisions are made by consensus. This does not mean that every one will be completely happy, but they every one can live with the decision and feel decisions are fair.
- No single method will work in every watershed. So depending upon the local soil natural resources agencies. Extension etc. Own methods of watersheds should be developed.
- The role of government should be shifted from a service provider to that of a facilitator for fulfilling the designed level of service on a sustainable basis. This can be realised by empowering the grassroots institutions and encouraging participatory management.

Conclusion

Every one agrees that the world is facing a severe water crisis. Water abundant regions have become water scarce and water scarce regions face water famines. Water crisis is an ecological crisis with commercial cause, but no market solutions. Market solutions destroy the earth and aggravate inequality. The solution to an ecological crisis is ecological and the solution for injustice is democracy. Ending the water crisis requires rejuvenating ecological democracy. Therefore, there is need for people's participation for sustainable development. Bottom-up approach is needed so as to involve local people in planning managing and decision making process. The bureaucracy should play the role of a facilitator. The maximum involvement of people, PRIs, NGOs, SHGs and other social groups may be encouraged by the Central and State governments.

REFERENCES

1. A.P.J. Abdul Kalam: Integrated Water Mission, Yojana, Vol. 4, June 2005.
2. Das Deepak Kumar: *"Environmental Impact of Inter Basin Water Transfer Projects"* Economic and Political Weekly, Vol. XLI, No. 17.
3. Dugal V.K: *Towards Safe Drinking Water and Sanitation*, Employment News, Vol. XXIX, No. 51.
4. Dreze, Jean and A. Sen (1995), *India: Economic Development and Social Opportunity*, Oxford University Press, Delhi.
5. Nair Pradeep: *Practising Participatory Communication for Development*, Employment News, Vol. XXX, No. 32.
6. Nathan K.K.: Hydrological Drought, Employment News, Vol. XXX, No. 32.
7. Nazrul Islam: *IRLP on the Ecological Approach to Rivers* EPW, Vol. XLI, No. 17.
8. Nimbal Farrok & Others: *"Rain Water Harvesting*, Yojana, Vol. 49, June 2005.
9. Petrella Riccardo: *Poverty, Water and Globalisation*, Yojana, Vol. 49, June 2005.
10. Pradhan K.C.: "Tribal Development and the Provision of PESA" in Dr. S.N. Tripathy, 2006, Edited Volume *"Rural Development" for Social Change*, Sonali Publications, New Delhi.
11. Sandra, Postel (1997), *Changing the Course of International Water Management*, Natural Resources Forum, Vol. 21, No. 2.
12. Shiva Vandana: *"The First and Last Commons"* Yojana, Vol. 49, June 2005.
13. Sharma Mukul: *The Making of Monal Authority*, Anna Hajare & WHP in Religion Siddhi EPW, Vol. XLI, No. 20.
14. Sharma Dr. S.K., *Watershed Management and Grassroot Institutions*, Kurukshetra, Vol. 54, No. 2.
15. Tiwari Rakesh: *"Conflicts Over International Water"* EPW, Vol. XIL, No. 17.
16. Verghese B.G. (1990): *Water of Hope*, Oxford and IBH Publishing Co., New Delhi.
17. White A, Shirly: *"Participatory Communication"*, 1994, Sage Publication, New Delhi.
18. World Commission on Dams (2000), *Dams and Development: A New Framework for Decision Making* (Report of the World Commission on Dames) Earthscan Publication, London.

15

Employment Generation Through Development of Tourism

*Virendra Thakore**
*Achintya Mahapatra***

Presiding over the meeting of the Parliamentary Consultative Committee of the Ministry of Tourism which was held at Bhubaneswar (Orissa) the Minister of Tourism and Culture, Smt. Ambika Soni remarked that the objective of the existing Tourism Policy of the Government is to position tourism as a major engine of economic growth and to harness its direct as well as multiplier effects for creation of employment and poverty eradication in an environmentally sustainable manner. The efforts made by the Ministry of Tourism and Culture in the last few years have had a salutary effect on the India Tourism. While the foreign tourist arrivals are expected to witness a growth of 78 per cent in 2006 over 2001 (last 5 years), growth in the foreign exchange earnings is expected to be of the order of 122 per cent during this period. India's share in world tourism, which was 0.37 per cent during 2001 end, is likely to increase to 0.53 per cent during 2006, according to Smt. Soni. Further, the Minister added that as per the estimates, on an average, about 3.1 million additional jobs per year have been created

* **Assistant Professor.**

** **Lecturer in Hotel and Tourism Management, Vivekananda Institute of Hotel and Tourism Management, Rajkot–360 003, Gujarat, <achintya4you@rediffmail.com>**

directly and indirectly in the tourism sector in the last 4 years. Smt. Soni underlined that tourism is a priority sector because, among others, it has the potentiality of maximizing the productivity of India's natural, human, cultural and technical resources. Tourism sector is labour intensive and cottage or small industry based providing employment; capable of being primarily focused on rural areas with appropriate and relatively low cost programmes and has extensive forward and backward economic linkages that build overall income and employment, especially for women and youth.

Giving details of her Ministry's initiatives in positioning and maintaining Tourism Development as a National Priority Activity, Smt. Soni said that a concerted effort has been made through the National Development Council (NDC) and other important forums by emphasising that tourism is now considered the world over a major engine of economic growth and several countries of the world have transformed their economies using the tourism potential to the fullest. Other initiatives include the setting up of National Tourism Advisory Council as a 'think tank', the constitution of a Group of Ministers on Tourism to improve policy integration and co-ordination. The development of a Tourism Satellite Account (TSA) to gauge precisely the contribution of tourism to the national economy using the methodology has been given by UN-World Tourism Organisation.

UN Statistical Division has provided an authentic tool to assess the success of the strategies. With a view to enhance India's competitiveness as a tourist destination, major policy initiatives such as liberalization in aviation sector; pricing policy for aviation turbine fuel which influences internal air fares; rationalization in tax rates in the hospitality sector; tourist friendly visa regime; immigration services, and; procedural changes in making available land for construction of hotels, allowing setting up of Guest Houses, etc. have been undertaken. These do not come directly under the purview of the Ministry of Tourism, but under the other Ministries and State Governments, have also been taken up vigorously, Smt. Soni said.

Smt. Soni said that the proposal to introduce Advance Passenger Information System (APIS) at airports to reduce delays in clearance of passengers at the immigration counters has already been accepted by the Ministry of Home Affairs and notified the system is being tested at Delhi Airport at present. This system, as and when introduced, will not only reduce delays in clearance but would also facilitate introduction of the scheme of visa-on-arrival. With a view to promote India as a safe destination, all the States have been requested to deploy tourist police at major tourist destinations in the States. The states of Andhra Pradesh, Karnataka, Goa, Kerala, Maharashtra, Himachal Pradesh, Rajasthan, J & K, Uttar Pradesh and Delhi have already deployed such police force. Matter has been taken up with Ministry of Home Affairs for issuing decisions to other States, she added.

A number of steps have also been taken to improve and expand the existing tourism products and their development. For expanding cultural and heritage tourism, integrated development of areas around these monuments is a prime activity. A few of the cultural and heritage sites/monuments taken up for development include—Red Fort, Humayun's Tomb and Qutub Minar in Delhi; Somnath Temple in Gujarat; Heritage properties at Chail and Naggar in HP; Hampi in Karnataka; Gwalior fort in MP; Ajanta Caves and Sinhagad Fort in Maharashtra; Nahargarh Fort, Amber Fort and Hawa Mahal in Rajasthan; Somnath and Taj Mahal in UP etc. Illumination of monuments has also been taken up in a big way, and financial assistance is being provided for the purpose, Smt. Soni said. A number of steps for the development of beach and coastal tourism have been taken. A number of sites on the west coast of India were identified for the development as beach resorts by the private sector. The sites on the beaches of Goa, Kerala and North Karnataka were taken up because of their easier access by air. Kochi in Kearla and the Andaman and Nicobar Islands were developed as international cruise destinations because of their proximity to international cruise routes and their exotic appeal. Smt. Soni informed that Rural tourism is being promoted as the primary tourism product of India to showcase rural like, art, culture and heritage at rural locates and in villages which have core competence in terms of crafts/handlooms/textiles/cultures, etc. to tourists, thereby also spreading tourism and its socio-economic benefits to rural areas.

Smt. Soni said that steps have also been taken to promote wildlife tourism, adventure tourism, MICE (Meetings, conventions and Exhibitions) tourism. In this context the Minister underlined the importance of developing world class international convention centres not just for the sake of tourism development but also for international domestic trade and commerce and added that presently, cities of Delhi, Mumbai, Agra, Bangalore, Cochin, Goa, Hyderabad, Jaipur and Kolkata are being promoted as MICE destinations in India but facilities available today are not enough. Smt. Soni underlined that India's most unique tourism product in the last five years has been holistic healing and rejuvenation packages. She added that in focussing on this, the essence of Indian culture has been captured both for international and Indian visitors alike. States of Kerala, Karnataka and Uttranchal have taken up Ayurveda and traditional techniques of healing as also rejuvenation therapy in a big way.

Smt. Soni also underlined the need for physical infrastructure for tourism which ranges from ports of entry to modes of transport (airways, roadways, railways or waterways) to urban infrastructure such as access roads, power, water supply, sewage and telecommunication. This undercrores the need for inter-sectoral convergence of infrastructural schemes and programmes that could support tourist destinations, she underlined. For effective marketing of India tourism, a number of new initiates have been taken in the last few years. These include direct approach to the consumers through electronic and print media, joint sales efforts with the international airlines operating to India and direct cooperative marketing with the tour operators and wholesalers overseas.

Ministry of Tourism is encouraging participation in travel marts like ITB, WTM, ATM, PATA etc. for promotion of India as a preferred tourist destination. These activities are in addition to the regular activities, undertaken by the overseas offices. A greater focus is being attached to the regions of China, North East Asia and South East Asia, Smt. Soni added and said that production of world class publicity material ranging from tourist information brochures, posters, scrolls, etc. with a common theme, that is, 'Incredible India' has been undertaken. The members of the consultative committee also visited Konark, which is a world heritage site and the Jagganath Temple at Puri. They also had visits to some other monuments in and around Bhubaneswar like Lingaraj temple, Rajarani temple, Mukteshwar and Khandgiri-Udaigiri caves etc.

In ordinary parlance unemployment is a situation of individuals' joblessness at the prevailing wage rate. Lord Keynes expresses that unemployment is the result of deficiency in effective demand in advanced economics. Machine becomes ideal as a result of deficiency in effective demand. In consequence of machines idealness demand for labour falls. It implies that there exists a positive correlation between effective demand and employment. According to Lord Keynes, unemployment may be arrested by keeping the level of effective demand sufficiently. This type of unemployment shown in the advanced countries is termed as cyclical or frictional unemployment by Keynes. In developing or underdeveloped economy like India, cyclical or fractional unemployment is not more serious. Such type of countries are suffering from the problem of chronic under-employment or disguised unemployment in the rural sector and also suffering from the urban unemployment among the educated class. It should be noted that unemployment problem in under developed or developing countries is not the consequence of deficiency of effective demand but due to shortage of the economy should create higher rate of saving for capital formation by managing resources tactfully.

Scenario of South Asia and India

Most of the countries of South Asia like India, Pakistan, Bangladesh, Bhutan, Nepal, and Sri Lanka are passing through the second stage of demographic transition. In this stage countries must suffer from the problem of population explosion, unemployment and slow rate of economic growth. Out of 1.4 billion people of South Asia, 60 per cent are in the working age group. Only 66 per cent of the total age group participates as labour force. Employment growth rate are lower than GDP and labour force growth rates. Agriculture is the main occupation of the people, although it suffers from the lack of investment and low productivity. One third of the total population of South Asia is in poverty. Illiteracy in South Asia is also a matter of great concern. In South Asia major portion of the total unemployed is youth.

In the year 1997, 70 per cent the total unemployment of Sri Lanka is youth 58 per cent of the total unemployment of India is youth and 45 per cent of total unemployment of Pakistan is youth. Declining trend of unemployment is public sector during the past decade dampens the prospects of employment of educated youth. Employment provided in the private sector is mostly in the semi-skilled or low-skilled section. In the year 1997, 41 per cent of the total higher secondary passed youth of India was unemployed. In South Asia women have limited opportunity to get themselves employed. Two third of South Asia women are engaged in agriculture and agriculture related activities. Women from the majority in the informal sector, is very advance although they are failed to get reasonable opportunity and wages. In the year 1999-2000 male and female unemployment rate in south Asian countries shown in the Table 15.1. The table shows the highest unemployment rate (16.2%) of women in Sri Lanka and lowest in Nepal (0.7%). In case of male unemployment, Sri Lankan rate (7.1%) is highest and Nepal is lowest (1.5%). Table 15.2 shows over all unemployment rates of South Asian countries in the year 1999-2000. The highest rate of unemployment is 8.2 per cent in Sri Lanka and lowest in Nepal, 1.1 per cent. Unemployment rate in India is 7.3 per cent in the year 1999-2000.

Table 15.1: Female and male unemployment rates in South Asia, 1999-2000

Countries	*Females (%)*	*Male (%)*
Bangladesh	2.3	2.7
India	10	7
Nepal	0.7	1.5
Pakistan	14.9	4.2
Sri Lanka	16.2	7.1

South Asia's unemployment level is increased from 2.9 per cent to 3.4 per cent during the period 1995 to 2001. The unemployment rate of India has risen from 6 per cent in 1994 to 7.32 per cent in 2001 (The Economic Survey of India). W.T.O. has failed to generate more employment in South Asia because of the unilateral trade measures adopted by the developed countries.

Table 15.2: Unemployment rates in South Asia (1999-2000)

Countries	*Rater (%)*
Bhutan	1.4
Bangladesh	3.3
India	7.3
Nepal	1.1
Pakistan	7.8
Sri Lanka	8.2

India which is a thickly populated country in South Asia, has a very large labour force. The quality and productivity of the labour force is poor and low. Only 8 per cent of the total work force is engaged in the organised sector and 92 per cent is engaged in the unorganised sector. Work force of the organised sector contributes near 40 per cent of national output. The status of the labour in the organised sector is reasonably good but the working condition in the unorganised sector. The fastest growing industry in India is it. In the year 2002-03 it earns revenues of Rs. 11,700 crore. In the year 2004 the industry alone created 1.85 lakh jobs in Karnataka. Though the majority of the jobs are in Bangalore, the largest number of jobs were lost also Bangalore. During the post liberalisation and globalisation period, the number of jobless workers in Karnataka is about 3 lakh. The number of closed state enterprises in Karnataka is 27.65 per cent of workers become contract workers in Mumbai. Employees of public sectors are forced to take VRS. The situation which implies the job security of workers is losing in India. Massive migration of labourers from rural areas to urban area is a serious problem of the Indian economy. 60 per cent of the rural labourers of Orissa have migrated due to loss of jobs.

Extent of Unemployment in India

A projection of national sample survey (NSS) on unemployment reveals that 16 million persons were unemployed and 12 million persons were estimated to be severely underemployed during the year 1990. It implies that the total backlog of unemployed persons in the beginning of the Eighth plan period (1992-1997) is 28 million (Table 15.3). The net additions to the labour force during 1990-1995 had been expected to be 37 million. So the total number of jobless persons during the eighth plan period (1992-1997) would be around 65 million. It is further expected that in the next five years, that is during 1995-2000, labour force would increase by 41 million. Hence by the year 2000 AD, the total number of job seekers would be around 106 million. The Planning Commission, therefore, concludes, "Employment growth in the aggregate over the estimated employment of 300 million in 1990 would have to be about 4 per cent compounded per annum if the goal of providing employment for all is to be achieved by the end of the Eighth plan and over 3 per cent per annum if it is to be attained by 2000 AD.

Table 15.3: Projection of unemployment for (1990-2000)

		Millions
1.	Backlog of employed in the beginning of 1990	28
2.	New entrants to the labour force during 1990-1995	37
	Total unemployed for the 8th plan (1 + 2)	65
3.	New entrants to the labour force during 1995-2000	41
	Total unemployed for the 9th plan	106

Source: Compiled from the Planning Commission, 8th Plan and 9th Plan.

Table 15.4 shows the year wise registration, placement, modified vacancies, submission made and live registers during the period 1994 to 2001. The number of unemployed persons (Live Register) is increasing very year during the period 1994 to 2001. It was lowest in 1994 (3,66,91500) and highest in 2001 (4,19,95900). The nature of year-wise registration during the study period is volatile. But on an average 55 lakh to 60 lakh unemployed persons register their names in each year of study as job seekers. In the year 1997 maximum unemployed persons, about 63 lakh registered their names with various employment exchanges of India. The placement was maximum (2,75,000) in the year 1997. But after the year 1997 number of placement was decreased gradually in each year of study. In the year 1996, number of notified vacancies was maximum (94,23,900). But after the year 1996 it decreased every year except the year the year 2001. 'Submission made' in the year 1994 was 37,23,400. In the year 1997 submission was 37,67,800 and after the year 1997 submission decreased in each year of study gradually. The above discussion expresses the gravity and extent of unemployment in India.

Table 15.4: Job-seekers of all categories registered with the employment exchanges in India

(Figure in Thousands)

Year	*Registration*	*Placement*	*Vacancies*	*Submission*	*Live Registration*	*% of P.I/Vac.*	*% of Pl./Sub*	*% increase in L.R. over previous year*
1994	5927.3	204.9	396.4	3723.4	36691.5	51.7	5.5	1.1
1995	5858.1	214.9	385.7	3569.9	36742.3	55.7	6	0.1
1996	5872.4	233.0	423.9	3605.9	37429.6	55	6.5	1.9
1997	6321.9	275.0	393.0	3767.8	39139.9	70	7.3	4.6
1998	5851.8	233.3	358.8	3036.6	40089.6	65	7.6	2.4
1999	5966.0	221.3	328.9	2653.2	40371.4	67.3	8.3	0.7
2000	6041.9	177.7	284.5	2322.8	41343.6	62.5	7.7	2.4
2001	5552.6	169.2	304.1	1908.8	41995.9	55.6	8.9	1.6

Source: DGE & T, Ministry of Labour, Govt. of India, New Delhi.

Tourism Industry and Employment Generation

At present, Tourism is one of the fastest growing industries in the world. It has become one of the most significant industries across the world. Now its position in respect of world is just below the oil and automobile industries. Tourism industry creates sufficient employment opportunity and brings foreign currency. To enhance national income, tourism plays a vital role. National integration and national understanding is maintained by the tourism industry. It plays a vital role in economic development of a nation, socio-culture and preservation of heritage. Tourism is an extremely labour-intensive industry with high linkage effect. As a world's largest export industry, tourism accounts for about 8 per cent of the total world's largest export, more than 30 per cent of the International trade in services, 10.7 per cent of the global work force and about 12 per cent of world GDP. According to the World Tourism Organisation (W.T.O.), in the year 2020 tourism will be an industry earning five billion dollars a day. The economic of the countries like Hong Kong, Singapore, Sri Lanka and small islands like Mauritius, Fiji, Malta largely depend directly or indirectly on this industry.

India is the sixth largest country in the world after China, Canada, Brazil, USA and Australia. The country is the mother of two primitive civilizations of the world. On is the Indus Valley Civilization and the other is the Aryan Civilization. India with her vast area with her physical size and appearances, cultural heritage, splendid natural resources, coastlines, mineral, forest cover, old and antique buildings, Tajmahal, one of the seven wonders of world, monuments, beautiful temples and mosques, forts, hill stations, scenic beauty, springs, rivers, canals, ponds, dams, flora and fauna, gardens, historical places, wild life sanctuaries can easily find a permanent place in the firmament of the tourism industry. The Government of India recognized tourism as an industry in the year 1967, establishing a department called 'Tourism and Civilization Aviation Department'. From this year the development of tourism in India is started. The Government of India declared the year 1991 as a year of tourism, to give more impetus on tourism development. A report of the Ministry of Tourism and Civil Aviation reveals that India's gross foreign exchange earning from tourism was Rs. 283 crores in the year 1997. In this year India received 19.9 per cent more tourists than the previous year. Share of business still confines to only about one per cent of the world's total tourist trade. In the year 1950 the total arrival of tourists in India was 15000. Up to the year 2004 the figure reached in 2.98 million. Now tourism industry is the third largest foreign exchange earned by producing gems and jewellery and readymade garments in India. Most of tourist coming to India belonging to countries— UK, USA, Sri Lanka, Germany, France, Japan, Canada, Italy, Malaysia and Singapore.

The role of tourism industry in generation of employment is very significant. Tourism is a labour intensive industry. So a large number of job seekers can get employment in this industry. According to a report from the Ministry of Tourism, Government of India, number of people already employed in the industry, is more than seven million. The number is about 2.4 per cent of Indian total labour force. People directly or indirectly depends on tourism industry in India is around eleven million. Tourism industry creates employment especially for hoteliers, restaurant owners, guides, local shopkeepers, merchants etc. The remarkable feature of the industry is that it employs a large number of both educated and uneducated women. Both skilled and unskilled women are employed in this industry. Women, in fact, out number men in hotels, airline services, travel agencies, handicraft and cultural activities are playing the major role in this industry. The estimated employment multiplier figure in the tourism industry is 2.36. It means the direct employment of 100 persons in tourism industry which create job for 136 persons in other sectors of our economy. From the view point of employment generation, this multiplier is very attractive than other industry-multiplier.

Conclusion

Unemployment problem in India is a mater of great concern like poverty, illiteracy, malnutrition etc. The nature of the problem is severing in a developing country like India. In India the number of unemployed persons is increasing constantly. Organised sectors fail to create sufficient employment opportunity to absorb the unemployment labour force. After independence, Government of India has taken many steps to tackle the problem. But these efforts have failed to solve the alarming problem of unemployment because the main cause behind this problem in India is population explosion. In this context, tourism industry can play a vital role in Indian economy. Tourism industry is a labour-intensive industry and its employment multiplier is 2.36. A large number of unemployed persons can be employed by this industry. The Government should take necessary steps for development of tourism industry. The Central Government and State Governments should allocate huge budgetary allocations for tourism industry. And the cooperation and co-ordination on the part of the Central and State Governments, local authorities and local people is needed for the development of the industry. The development of tourism industry mainly depends on adequate hotel accommodation, better transport facilities, medical and communication facilities etc. The Government should take necessary steps to improve these facilities for the growth of tourism industry in India. The maintenance of safety (like from terrorism and naxalite) in the destination is also important points which should be checked by the local Government. The beggars are creating lot of problems to the outsiders at all tourist spots. The local cheaters like taxi, auto drivers, rickshaw pullers, shops, small hotels and restaurants owners should also be checked.

REFERENCES

1. Konar Dr. D., "*The Scenario of Population Growth in India*" Akansha Publishing House, New Delhi, 1st Edition, 2004.
2. *Economic Survey*, Govt. of India, Various Issues.
3. Chakrabarty, Amit Kumar., *JOHAR—Journal of Hospitality Application and Research*" Vol. 1, No. 2, July 2006.
4. Tripathy Dr. S.N., "*Rural Development for Social Change*" Sonali Publications, New Delhi–110 002, 2006.
5. Methani, M.C. *Tourism*, Yojana, Vol. 35, No. 8, May 15, 1991.
6. *Website of the Ministry of Tourism*, Government of India-2006.

Index